SCIENCE AND RELIGION
IN ARCHAIC GREECE

Roger Sworder

SCIENCE & RELIGION IN ARCHAIC GREECE

Homer on Immortality
&
Parmenides at Delphi

SOPHIA PERENNIS

SAN RAFAEL, CA

First published in the USA
by Sophia Perennis
© Roger Sworder 2008

Series editor: James R. Wetmore

For information, address:
Sophia Perennis, P.O. Box 151011
San Rafael, CA 94915
sophiaperennis.com

Library of Congress Cataloging-in-Publication Data

Sworder, Roger.
Science and religion in archaic Greece / Roger Sworder. — 1st ed.

p. cm.

Includes bibliographical references.
ISBN 978-1-59731-087-1 (pbk: alk. paper)
1. Religion and science—Greece. 2. Greece—Religion.
3. Science—Greece. 4. Homer—Religion
5. Parmenides. I. Title
BL240.3.S96 2008
292.1'650938—dc22   2008022418

CONTENTS

ACKNOWLEDGMENTS

The author would like to acknowledge The Prometheus Trust, T. J. Addey, Chairman, 194 The Butts, Frome, Somerset, UK, for their publication of the first edition of *Homer on Immortality*, and their gracious permission to permit its inclusion in the present volume. He would also like to express his gratitude to Dorothy Avery, Rod Blackhirst, and Cliff Carrington for their help and advice in the preparation of this book.

HOMER ON IMMORTALITY

If I knew Valpy, I should certainly expostulate with him for allowing Taylor the Platonist to write in his journal. The man is an ass, in the first place; secondly, he knows nothing of the religion of which he is so great a fool as to profess himself a votary; and, thirdly, he knows less than nothing of the language about which he is continually writing.

—The Literary Idler, *Blackwood's Magazine*, June, 1825

INTRODUCTION

I HAD STUDIED Latin, Greek and Classical Civilization at primary, secondary and tertiary levels, but it was not until my late twenties that I began to take Homeric religion seriously. At that time I read Thomas Taylor's translation of Porphyry's essay *Concerning the Cave of the Nymphs* and his translation of Proclus' *Apology for the Fables of Homer.* Thomas Taylor, I learnt then, had lived in London two centuries ago. As an amateur, with no professional qualifications as a scholar, he had devoted himself to the study of Homer, and to the study and translation of Plato and the Neoplatonists. In England his work met with little general success and with active hostility from many. But William Blake knew Taylor and was influenced by him in his treatment and perhaps choice of the Cave of the Nymphs as a subject for his painting. Coleridge and Keats also refer to Taylor and the effect of his work upon them. Through these men, if not through his own work directly, Taylor was a major figure in the English Counter-Enlightenment. For the rest he was, like Blake and Coleridge, a notable eccentric of his time, whose wholehearted enthusiasm for ancient Greece as a source of spiritual wisdom marked him out from his fellow scholars then and now.

For Taylor, as for Proclus and Porphyry, each appearance of a God or Goddess in the Homeric epics is a revelation, a depiction of an eternal principle under the guise of fable. At their highest, Taylor thought, these principles belonged to the foundations of mathematics and logic, and in this he followed Proclus, the Neopythagoreans and the Neoplatonists. It was as a mathematician that Taylor had first come to a study of Greek philosophy. Taylor was criticized for his dependence on these later sources and accused of anachronism on the ground that commentators on Homer in the early Christian period could have known little of the poet's intentions a thousand years earlier. The same criticism was made of Taylor's writing on Plato. But whether or not Taylor was correct to rely on these later sources, his attempt to study the Homeric poems as primarily religious works had the warrant of a very much earlier authority, the historian Herodotus:

> But it was only—if I may so put it—the day before yesterday that the Greeks came to know the origin and form of the various gods, and whether or not all of them had always existed; for Homer and Hesiod, the poets who composed our theogonies and described the gods for us, giving them all their appropriate titles, offices, and powers, lived, as I believe, not more than four hundred years ago.[1]

Since the *Iliad* and *Odyssey* comprise by far the greater part of the combined works of Hesiod and Homer on any reckoning, it is clear that according to Herodotus these epics were significant in the establishment of the forms and nomenclature of ancient Greek religion. Yet it is rare to see these poems treated as theological in their own right, though critiques of the Neoplatonic theology attributed to Homer by later theorists such as Proclus are not uncommon.

Between Taylor's time and our own there has developed the method of comparative anthropology to study ancient religions. The founder of the method was the classicist Sir James Frazer and he applied his method to ancient Greek religion. Frazer was more interested in descriptions of ancient rituals and religious sites than he was in Homer's theophanies, and he assimilated Greek religious practices to his knowledge of what he called primitive customs still extant around the world. For Frazer and even more for Jane Harrison who followed him, the Gods whom the Greeks worshipped in their rituals with superstitious fear had little in common with Homer's Gods. For Harrison, Homer's Gods were detached from their roots in the living practice of their time like 'a bouquet of cut flowers.'[2] This conception of the Homeric pantheon fits it perfectly for those baroque Classical confections which graced the seventeenth and eighteenth century courts of Europe. And Homer does lend himself to this. It is hard for us to take seriously Gods who misbehave as flagrantly and as comically as Homer's do. Plato thought

1. Herodotus, *The Histories*, tr. Aubrey de Sélincourt, rev. by A.R. Burn (Middlesex: Penguin Books Ltd., 1972), II, 53.

2. Jane Harrison, *Themis* (London: Merlin Press, 1963), p xi.

Homer's treatment of the Gods open to abuse and the case of Jane Harrison bears him out.

It is odd that the study of ancient Greek religion should have fallen under the discipline of anthropology and not of theology. Should the study of ancient Greek religion, or indeed of any religion, be part of the study of man rather than of God?

According to Ludwig Wittgenstein, Frazer had almost no religious sensibility:

> Frazer is much more savage than most of his savages, because these won't be as far from understanding spiritual matters as an English man of the twentieth century. His explanations of primitive customs are much cruder than the meaning of these customs themselves.[1]

Other thinkers rejected the anthropological approach but used the comparative method on the same material to a quite different end. They assimilated the Homeric theophanies to scriptures from around the world in their account of universal spiritual themes. In the pages of René Guénon and Ananda Coomaraswamy, Homer's epics serve as adequately to illustrate spiritual matters as the Vedas, the Chuang Tze or the Buddhist Sutras. Guénon's brief essays *The Solstitial Gates* and *The Symbolism of the Zodiac among the Pythagoreans* discuss Porphyry's *Concerning the Cave of the Nymphs*,[2] and I draw on them in what follows. But these accounts of Homer are *en passant* and piecemeal. The task of understanding the Homeric theology systematically within the context of his epics remains to be done. Taylor is yet to find a successor. Homer is hardly at all studied now as the religious teacher whom Herodotus acknowledged. Even the new discipline of religious studies is naturally more concerned with living traditions.

The ancient Greeks supposed Homer to be a great religious teacher; most modern scholars do not. The ancient Greeks believed Homer to be a great scientist and most modern scholars do not. According to Strabo who lived in the first century BC, Homer was

---

1. 'Vemerkungen über Frazer's *The Golden Bough*', *Synthese*, ed. Rush Rhees, 1967, vol. 17, pp 233–253.

2. René Guénon, *Symbols of Sacred Science* (Hillsdale, NY: Sophia Perennis, 2004).

the founder of the science of geography[1] and had known that the earth was round.[2] Strabo's *Geography* is the first extended work in the science of geography to have survived to us from the ancient world. A century before Strabo, Crates the grammarian, librarian of the great library at Pergamum, manufactured a wicker sphere in order to illustrate Homer's account of the journeys of Odysseus.[3] According to Crates, Homer describes at least the arctic and antarctic zones and the southern tropic in the *Odyssey*. Crates maintained, perhaps even began, a tradition in Homeric interpretation to which Porphyry belonged several centuries later. According to Porphyry, the gates to the Cave of the Nymphs in the *Odyssey* are the tropics of Cancer and Capricorn.[4]

The ascription of such knowledge to Homer defies modern histories of science as to how Greek science developed. Or perhaps, more reasonably, the accepted view of how Greek science developed defies the testimony of Crates, Strabo and Porphyry. According to Sir Thomas Heath, Homer knew of the sun and moon, the morning and evening stars, the Pleiades, Hyades, Orion, the Great Bear, Sirius, Arcturus, and Bootes; he thought the earth was flat and surrounded by the River Ocean; and he made vague use of astronomical phenomena to fix localities and to tell the time of day:

> So far as Greece is concerned, astronomy as a science, in however elementary a stage, begins with the Ionian philosophers, with whom Greek philosophy and Greek mathematics also begin.[5]

But on the view of Crates, Strabo, and Porphyry, Homer established a complex astrogeography, which determined the regions of the earth in relation to the movements of the stars, and especially to the yearly course of the sun, in a comprehensive and systematic way. Furthermore, on the view of those ancients, this Homeric astrogeography is woven into the metaphor of the poems and is integral

---

1. Strabo, *Geography*, tr. H.L. Jones (London: Loeb Classical Library, 1969), I, 1.2.
2. Ibid., I, 1.11.
3. Ibid., I, 1.20.
4. *Thomas Taylor the Platonist*, eds. Kathleen Raine and George Harper (Princeton: Princeton University Press, 1969), pp 309ff.
5. Sir Thomas Heath, *Greek Astronomy* (New York: Dover Publns, 1991), p xii.

to them. The astrogeographical data are not presented as discoveries but as profoundly meditated symbols. It may be that Crates, Strabo and Porphyry are wrong about Homer's knowledge and that our historians of science are right. But this can only be ascertained after a lengthy and detailed analysis of the Homeric epics. If it turns out that Crates and his followers had warrant for their readings in the epics themselves, then the history of Greek science will need some revision. But of course we are dealing with poems, not research papers, and Homer's knowledge, even if it is there, may take some recognizing.

In the account of the *Odyssey* which follows I have set myself two tasks: to reassess Homer's achievement or lack of it as a spiritual teacher; and to reassess Homer's achievement or lack of it as a scientist. If we examine Porphyry, the major exponent of Homeric symbology whose work has survived to us, we find that these two enquiries are very closely linked. According to Porphyry, Homer's astrogeography is one of the major sources of symbol in his religious teaching. On that view, we cannot assess Homer as a religious teacher unless we examine his achievement as a scientist. Can we assess him as a scientist without examining his religious teaching? On the modern understanding of science this should certainly be possible, even necessary, if we are to have a clear view of Homer's claim to what we would call scientific knowledge. But as Thomas Taylor says again and again, Enlightenment Science wholly fails to recognize the dependence of material phenomena upon divine or spiritual principles. That is why, in his view, we know more than we ever wanted to know about how things work and almost nothing about why they should exist in the first place. So the question of whether we can assess Homer as a scientist without assessing him as a religious teacher is a hotly contested one.

I have said enough to give the reader some idea of what to expect in the following pages. The matter may be put even more briefly. My account of the *Odyssey* differs from most others because it adopts two postulates for examination, neither of which is widely adopted. Homer was a great religious teacher; Homer was an accomplished astrogeographer. Of course the view of the *Odyssey* which emerges from an examination of these postulates is very different from those views of the *Odyssey* which currently prevail. I will finish with some words of Thomas Taylor which first moved me many years ago, the kind of writing which Valpy, God bless him, published in his journal:

> Such readers, too, as may fortunately possess a genius adapted for these speculations will find that the fables of Homer are replete with a theory no less grand than scientific, no less accurate than sublime; that they are truly the progeny of divine fury; are worthy to be ascribed to the Muses as their origin; are capable of exciting in those that understand them the most exalted conceptions, and of raising the imagination in conjunction with intellect, and thus purifying and illuminating its figured eye.[1]

1. *Thomas Taylor the Platonist*, op. cit., p449.

1

ODYSSEUS
IN THE ANTIPODES

WHERE EXACTLY did Odysseus go in the course of his journey from Troy to Ithaca? Where were Phaeacia and Ogygia, the Cimmerians, Aeaea and Laestrygonia? These questions provoked passionate debate in the ancient world as they have in the modern, and many places in the western Mediterranean have been claimed as the sites of Odysseus' adventures. The most favoured locations have been the western coast of Italy around Naples; but the straits of Messina between Italy and Sicily, and the island of Corfu, have also figured prominently in these reconstructions. Recently Mr. Tim Severin has sailed these seas in a replica of Odysseus' vessel and made a miraculous rediscovery of Scylla's cave on the shores of Arcania.[1] In ancient times Strabo devoted the first book of his *Geography* to an account of Homer's description of the earth. Strabo lived at about the time of Christ and drew on the works of the greatest geographers before him, such as Eratosthenes and Apollodorus, for his analysis of Homer's geography. Strabo wrote:

> Apollodorus, agreeing with Eratosthenes and his school, censures Callimachus, because, though a scholar, Callimachus names Gozo and Corfu as scenes of the wanderings of Odysseus in defiance of Homer's fundamental plan, which is to transfer to Oceanus the regions in which he describes the wanderings as taking place.[2]

Apollodorus and Eratosthenes before him supposed that the jour-

---

1. Severin, Tim, *The Ulysses Voyage* (London: Arrow Books Ltd., 1987), pp 226–232.
2. Strabo, op. cit., I, 2.37.

ney of Odysseus did not take its course through the known world at
all but out on the Ocean, and they objected to the identifications of
Callimachus on the ground that Homer had never intended to be
understood in this way. Strabo's own view is that the journey is not
entirely a fiction, but a mixture of myth with local details; provided
that none of the places claimed as sites of the adventures is more
plausible than those claimed by Callimachus, Callimachus' claims
must stand with the rest. These experts all lived much closer to
Homer's own time than we do, and if they were incapable of making
up their minds about these details, our chances are hardly likely to
be better. The fact is that Homer's descriptions in the *Odyssey* are
not enough to enable the making of an adequate itinerary. For
sometimes he tells us the days taken on a leg of a voyage, at other
times the direction, but rarely both and often neither. Circe's island
of Aeaea, from which Odysseus travels to the dead and to which he
then returns, and from which again he sets out on his journey to
Calypso, is very difficult to locate in relation to Laestrygonia from
which he first arrives at it or to Calypso's island for which he finally
departs from it. This same confusion is felt by Odysseus and his
companions in the story when they complain, on first arriving at
Aeaea, that they have lost all sense of direction and cannot even tell
where the sun rises and sets. Later Homer tells us that the island of
Aeaea is the dwelling and dancing place of the Dawn.

Faced with these difficulties Eratosthenes, Apollodorus and
Strabo too, seem to have cut the knot. They have all 'transferred to
Oceanus the regions in which Homer describes the wanderings as
taking place.' This means perhaps that whenever one of those
tedious questions arose as to where exactly Odysseus did this or
that, these experts would reply that the discussion was otiose since
the adventures took place somewhere out on the Ocean and that
was all that could safely be said. Homer and the Greeks understood
the Ocean to be a vast indefinite sea circumscribing the inhabited
world. The transference of the adventures to the Ocean may remind
us, then, of those early maps on which cartographers have filled up
little known areas with legends such as 'Here be monsters'. And
there are monsters enough on Odysseus' travels. It is also true that
Homer himself refers to the Ocean several times in his account of

Odysseus' adventures: Odysseus reaches the bank or shore of Ocean when he travels to and from the dead. On this evidence, then, the early geographers had some justification for transferring the adventures in this way. It was not merely a means to escape the many difficulties which a literal reading of the *Odyssey* entailed. But it is also clear that Apollodorus, and those who agreed with him, condemned anyone who attempted to locate Odysseus' adventures in the known world. Callimachus 'though a scholar' had been guilty of this lapse and had failed as a scholar in being so. Strabo is much more inclusive and acknowledges the existence of local references in Homer's descriptions.

But Strabo too transfers many of Odysseus' adventures to Ocean, and seems to have made up the terms Exoceanism to name the practice, as well as the verb 'to Exoceanize'. This Exoceanic way of reading Homer's account of Odysseus' adventures obviated the need for local identifications but may have been much more than just a washing of hands to avoid the whole problem. Many ancient geographers had a way of reading Homer's descriptions and directions which turned them into a kind of astrogeography, a systematic account of the earth's surface on the largest scale. These early geographers felt a great debt to Homer whom they considered the founding father of their science. This is how Strabo describes Homer in the first pages of his *Geography*. On this view Homer is not concerned so much to locate particular cities or places in his account of Odysseus' adventures as he is to determine geographical zones by means of seasonal variations, changes to the visibility of stars in the night sky at different latitudes, and by differences in the lengths of days and nights at the solstices. These were the methods employed by all ancient geographers, and in the view of Strabo and others they were all employed by Homer. Homer, according to Strabo, had gone to the uttermost bounds of the inhabited world, encompassing the whole of it in his descriptions.[1] In this way, then, it was possible to 'exoceanize' Homer's descriptions in a positive manner and understand his work as scientific rather than either fictional or literal.

1. Strabo, op. cit., i, 1.2

So let us too exoceanize along with Strabo and this group of interpreters and see where it takes us. We may begin at the very top with Homer's description of the Great Bear and the constellation Orion. At this point in the story, Odysseus is just setting off from Calypso's island towards Phaeacia.

> It was with a happy heart that the good Odysseus spread his sail to catch the wind and used his seamanship to keep his boat straight with the steering oar. There he sat and never closed his eyes in sleep, but kept them on the Pleiads, or watched Bootes slowly set, or the Great Bear, nicknamed the Wain, which always wheels round in the same place and looks across at Orion the Hunter with a wary eye. It was this constellation, the only one which never bathes in Ocean's Stream, that the wise goddess Calypso had told him to keep on his left hand as he made across the sea.[1]

The constellation of the Great Bear is usually taken as the one that marks the North Pole since it always closely circles the North Pole, and it never sinks beneath the horizon. Always visible in the night sky in much of the northern hemisphere if weather conditions permit, it is the only constellation according to Homer which never bathes in Ocean's stream. All the other stars rise in the east and sink in the west, but the Bear is, in this special way, imperishable. Orion, on the other hand, rises and sets, and for a certain period of the year cannot be seen at night. But Orion is also the largest of all the constellations and turns in its course along the celestial equator. In these ways the Great Bear and Orion mark the two key areas for an understanding of the movements of the heavens in the northern hemisphere: the Great Bear turns around the North Pole and Orion turns around the equator. In a single image Homer outlines the fundamental armature or structure which governs the movement of the fixed stars.

It is a terrible image. These two vast figures in the sky are fixed at this first stage of their combat as they circle around, caught in this moment of everlasting peril, the Bear with her back to the Pole. The tension between the Bear and the hunter stretches the universe on

---

1. Homer, *Odyssey*, tr. E. V. Rieu (Middlesex: Penguin Books, Ltd., 1946), v, pp 269–277.

its frame. For as long as the bear keeps her eye on the hunter she is safe, and so she turns and turns, never shifting her gaze, never giving Orion the opening to launch his attack. The Greeks can always see the Bear but they cannot always see Orion. But as they watch the Bear turning round the Pole, it is easy for them to imagine that she is keeping her eye on Orion even when Orion himself is not visible to them, being below the horizon. As they watch the Bear turn around the part of the sky on the other side of the Pole from where they stand, they can picture Orion still hunting since the Bear is clearly as vigilant as ever.

In Homer's image it is the Bear which is watching Orion, not Orion the Bear. Homer asks us to imagine the scene through the eyes of the Bear. Being set up there, next to the Pole, the Bear must have a very good view of the world and Homer invites us to share it for a moment. Where the Bear is, there is not much movement at all. At the still point of a turning world, she can watch the movements of the stars both above and below our horizons simultaneously. This point of view plays a part in many early traditions. The intellectual act of envisioning the movements of the heavenly bodies with the mind's eye has a kind of physical correlative in the act of standing at the North Pole and actually witnessing the movements of the stars all the way round the earth. The god associated with the far north in the Greek tradition was the god of astronomy, Apollo. In this connection we may recall the famous golden thigh of Pythagoras, which Pythagoras once showed to Abaris to prove to him that he, Pythagoras, was Apollo of the far north,[1] or at least his chosen follower. The thigh was a symbol of the Great Bear. In Egyptian temples the Great Bear was represented by the hind leg of a cow. It is apt that this symbol of the pivot of the world should be a sacrificial offering.

Homer does not mention the people of the far north, the Hyperboreans, in the *Iliad* or *Odyssey*. But he does talk of another far northern people by name, the Laestrygonians. Odysseus and his companions come to Laestrygonia after leaving Aeolus' island for the second time. Odysseus' encounter with the Laestrygonians leads to

---

1. *The Pythagorean Sourcebook and Library*, ed. K.S. Guthrie (Michigan: Phanes Press, 1987), Iamblichus, 'The Life of Pythagoras', p 80.

his losing eleven of his twelve ships and all their men. For the Laestrygonians like the Cyclopes are huge as the Giants and share with the Cyclopes a taste for human flesh. They pelt Odysseus' ships with rocks and spear his sailors in the water like fish. Odysseus and his shipmates escape in the one ship which did not go into the harbour.

In his *Elements of Astronomy*, written in the first century BC, Geminus quotes from Homer's description of Laestrygonia as he explains what we call latitudes:

> The lengths of days are not the same for all countries and cities. The days are longer for those who live towards the north and shorter for those who live towards the south. The longest day in Rhodes has 14½ equinoctial hours, the longest in Rome 15 equinoctial hours. For those farther north than the Propontis, the longest day has 16 hours, and for those still farther north, 17 and 18 hours.... Crates, the grammarian, observes that even Homer mentions these regions in the passage where Odysseus says: 'Telepylos of the Laestrygonians, where the herdsman driving in his herd calls to herdsman, and the herdsman driving out answers him. There an unsleeping man might earn the wages of two, one for tending cattle, one for pasturing white-fleeced flocks; for there the paths of night and day come near.
>
> For in these regions, the longest day having 23 equinoctial hours, the night is quite short, being left with only one hour, so that the setting of the sun is near to its rising, since only quite a small arc of the sum-mer-tropical circle is cut off under the horizon. If then, says the poet, any one could keep awake through days of this length, he could earn two wages, 'one for tending cattle, one for pasturing the white-fleeced sheep.' Then he adds the reason, which is mathematical and connected with the theory of a sphere, 'for the paths of night and day come near,' that is, the time of setting is close to the time of rising.[1]

Geminus attributes this way of understanding Homer's descrip-tion of Laestrygonia to Crates the grammarian, and endorses Crates' interpretation. This Crates, citizen of Mallus, ambassador of King Attalus to Rome, librarian to the great library at Pergamon, made a globe of the earth and illustrated Odysseus' journeys upon it. Crates

---

1. Heath, *Greek Astronomy*, op. cit., Geminus, 'Elementa astronomiae', 6, pp132–133.

lived a century or so before Geminus, in the second century BC. Strabo, in his *Geography*, also draws extensively on the work of Crates in his account of Homer.

It is characteristic of Homer that when he describes this extraordinary geographical phenomenon in Laestrygonia he should immediately examine it from the angle of economic advantage. This is the poet who in the *Iliad* touchingly describes an exchange of armor on the battlefield between family friends who are now on opposite sides, and then cynically concludes by pointing out that one set of armor was worth ten times the other and that the loser was a fool.[1] But we can see also from this passage about the Laestrygonians how Homer's location of them latitudinally, as we might say, allows for a description as poetic as it is exact. The locational device is an occasion for wit.

At St. Petersburg in Russia, the night is just fifty minutes long at midsummer, so that conditions there would be very like those Homer describes for Laestrygonia. Homer's lines describing day and night in Laestrygonia, are plainly due to some knowledge of the land of the midnight sun. It is difficult to determine how vague this knowledge was. The fact that days and nights become progressively more unequal in winter and summer as one goes north from the equator must have been a matter of common observation in Homer's time. It would have been easy to extrapolate from this that at the North Pole there were sometimes no days or nights in the usual sense. The first sea journey to the far north which we know to have been made by a Greek was made by Pytheas, the circumnavigator of Britain, about 300 BC.[2] Pytheas was a citizen of the Greek city of Marseilles which was founded about 600 BC. In the sixth century BC there is evidence of Greek architects working near Munich, using a walling technique to be found at Delphi from the same period.[3] We must not underestimate the reach of Greek exploration in the

---

1. Homer, *Iliad*, tr. E. V. Rieu (Middlesex, Penguin Books, Ltd., 1950), VI, 234–236.

2. C. Kaeppel, *Off the Beaten Track in the Classics* (Melbourne, Melbourne University Press, 1936), pp 113–139.

3. John Boardman, *The Greeks Overseas* (Middlesex, Pelican Books, Penguin Books, Ltd., 1964), p 229.

archaic period. But whether it is by extrapolation or by a more direct acquaintance, Homer describes the Laestrygonians as a people near the Pole.

This way of analyzing Homer's description of the Laestrygonians explains this adventure of Odysseus according to the positive doctrine of Exoceanism. I have compared the astrogeographical conditions which obtained in Laestrygonia to those of St. Petersburg, where at midsummer the nights are as short as Homer describes. But I am not claiming that Homer was referring by this description to some actual city of the far north when he told the tale of the Laestrygonians. The point of the description is not that it refers to an actual place to be located on the globe very near the North Pole. Homer's description refers only to a region. This region is somewhere in the far north, but is no more determinable than that. The conditions which Homer describes apply to the entire northern portion of the northern hemisphere, all the way round the arctic circle. His description is universal and systematic rather than local.

One other feature of the Laestrygonians is worth noticing: they are huge. Odysseus' advance party into Laestrygonia is appalled at the size of the queen, huge as the peak of a mountain, when they first enter her high-roofed palace. There was a general belief in the ancient Greek and Roman world that the peoples of the north were bigger than the peoples of the south such as themselves, that the northerners were taller and sturdier than southerners. Porphyry wrote in the second century CE:

> The southern regions produce small bodies; for it is usual with heat to attenuate them in the greatest degree. But all bodies generated in the north are large as is evident in the Celts, the Thracians and the Scythians; and these regions are moist and abound with pasture.[1]

On this view the size of the Laestrygonians agrees with their location in the far north, as does their pasturing way of life. But it is ironic that people who live close to the Pole, that symbol of the absolute, should be so savage and uncouth. Except for the Hyperboreans, proximity to the Pole does not seem to generate a very high

---

1. *Thomas Taylor the Platonist*, op. cit., pp316–317.

level of culture. Nor does it bring happiness if we are to judge from the other polar people whom Homer describes, the Cimmerians.

Odysseus reaches the Cimmerians on his way to the dead when he is sent there by Circe with a fair north wind to help him.

> All we had to do, after putting the tackle in order fore and aft, was to sit still, while the wind and the helmsman kept her straight. With a taut sail she forged ahead all day, till the sun went down and left her to pick her way through the darkness.

> Thus she bought us to the limits of the deep-flowing River of Ocean, where the fog-bound Cimmerians live in the City of Perpetual Mist. When the bright Sun climbs the sky and puts the stars to flight, no ray from him can penetrate to them, nor can he see them as he drops from heaven and sinks once more to earth. For dreadful Night has spread her mantle over the heads of that unhappy folk. Here we beached our boat. . . .[1]

Odysseus is going south when he reaches this place and he comes to the limit of Ocean. In Homer's description of the Great Bear, we have a constellation which does not bathe in Ocean's stream because it is so far north. The stars of the arctic circle in this way limit the Ocean and the same would be true of an antarctic circle. If we take it that this is what is meant when Homer describes Odysseus as going south to the limits of the Ocean, then there are other features of his account of the Cimmerians which confirm it. In the country of the Cimmerians, the sun's rays reach them neither during the sun's climb into the sky nor during its descent to the earth. These conditions are met during midwinter near the South Pole when for many days and nights there is only night and the sun does not appear above the horizon at all. On this reading the mist and cloud in which Homer describes the Cimmerians as wrapped become meteorological metaphors for the midwinter darkness near the Pole, and not the causes of the darkness.

In this way we may pair the Cimmerians with the Laestrygonians as the two peoples of the Poles. Odysseus' journey from one to the other may be seen as spanning the entire surface of the earth from

---

1. *Odyssey,* XI, 9–20.

north to south. His journey is exemplary, not just one journey among other possible journeys, but a long journey over almost the whole earth along this line. It may be that Laestrygonia is not as far north as Cimmeria is south since there is still some night in Laestrygonia but no day at all in Cimmeria. In fact, of course, the astrogeographical conditions are identical at the North and South Poles and they have exactly the same amount of light and dark as each other. Homer has described Laestrygonia at or near midsummer and Cimmeria in midwinter, but it could just as easily have been the other way round. It is interesting from this point of view to consider the length of Odysseus' stay with Circe on his first arrival at Aeaea. It is exactly one year. He came to Aeaea from Laestrygonia and he leaves Aeaea for Cimmeria. But if he left Laestrygonia in the northern midsummer and a year later visited the antarctic circle, it would be the southern midwinter there just as Homer has described it. This is, of course, to leave out of account the time it took for Odysseus to travel from Laestrygonia to Circe's island and from there to Cimmeria.

And we still have the problem that the South Pole is no darker than the North Pole and therefore no more appropriate as a setting for the dead. And there is the other question of how much Homer could have known about the South Pole in any case. To appreciate the phenomenon of the celestial South Pole one need go no further south than the equator, and Homer may have known of this. Crates of Mallus is recorded by Strabo as having said that Odysseus' sea journey to the dead took place on the waters of 'a kind of estuary or gulf which stretches from the winter tropic towards the South Pole.'[1] Crates was an accomplished Exoceanist and his testimony adds some weight to our reading of the Cimmerians. Or perhaps we may feel that Crates was guilty here of an anachronistic retrojection, imputing to Homer a knowledge of global conditions which Homer could not possibly have shared. Crates himself made a globe to represent the earth, and Strabo urges those who would do the same to make theirs at least ten feet in diameter.[2]

The name of Cimmeria was common in the ancient world and

1. Strabo, op. cit., I, 1.7.
2. Ibid., II, 5.10

Homer's use of it for a southern place in or near Hades is striking. The historical Cimmerians occupied the Bosphorus and invaded Asia Minor during Homer's lifetime according to some ancient writers. The Cimmerians were perhaps a direct threat to him as a citizen of that region. They were known for the gloominess of the country which they inhabited. On one account, recorded by Strabo:

> And, too, it was on the basis of Homer's actual knowledge that the Cimmerians lived about the Cimmerian Bosphorus, a gloomy country in the north, that he transferred them, quite appropriately, to a certain gloomy region in the neighborhood of Hades—a region that suited the purpose of his mythology in telling of the wanderings of Odysseus. The writers of chronicles make it plain that Homer knew the Cimmerians, in that they fix the date of the invasion of the Cimmerians either a short time before Homer, or else in Homer's own time.[1]

Strabo has serious reservations about this claim. If it were true, then Homer's location of the Cimmerians near Hades was apt enough. They lived in a gloomy place. They had invaded his country from the north and Homer merely wished them further in the same direction, with that special kind of poetic justice which sometimes gives poets the last word.

Having established Laestrygonia on the arctic circle and Cimmeria towards the South Pole, we may now examine the whereabouts of Circe's island Aeaea which Odysseus visits on his way from Laestrygonia to Cimmeria. Was Aeaea in the northern or southern hemisphere? According to Strabo the words spoken by Odysseus to his companions when they first arrive at Aeaea give the clue to its location:

> 'My friends', I said, 'East and West mean nothing to us here. Where the Sun is rising from whence he comes to light the world, and where he is sinking, we do not know. So the sooner we decide on a sensible plan the better—if one can still be found (which I doubt).'[2]

These words Strabo understands as indicating that they have

---

1. Ibid., 1, 2.9
2. *Odyssey*, x, 90–93.

travelled so far south that the stars of the familiar northern latitudes have disappeared, and that is why they cannot orient themselves.[1] If that is so, and they were near the arctic circle in their encounter with the Laestrygonians, we may conclude that they had left the northern hemisphere altogether when they came to Aeaea, and could no longer see any part of the Great Bear. So we may suppose on this account that Aeaea was in the southern hemisphere.

On the other hand Homer tells us that Odysseus arrives on Aeaea when it is hot. He kills a stag which is seeking water because it is thirsty from the heat. More significant still is one line in Homer's account of how long Odysseus stayed on this first visit.

> In fact we stayed on day after day for a whole year, feasting on meat galore and mellow wine. But when the year was out and the seasons began to repeat their round, as the months waned and the long days were brought in their course, my good friends called me aside one day and said reproachfully. . . .[2]

The crucial phrase here is 'and the long days were brought in their course' since this detail tells us that Odysseus left for Cimmeria during the Aeaean summer. But it was midwinter in Cimmeria. It is, of course, true that any dealings with Circe seem to occasion these violent oscillations between extremes, but the fact remains that if it was midwinter in Cimmeria when it was summer in Aeaea and Cimmeria is near the South Pole, then Aeaea must have been in the northern hemisphere. But this is the opposite conclusion to the one we reached a moment ago. It is interesting to note in this context that the line with the phrase 'and the long days were brought in their course' was omitted from many early copies. We may wonder whether it appeared in Strabo's edition of Homer, and whether its frequent omission is the result of just such considerations as we have discussed here, since in these considerations and perhaps no others this clause has a special significance. On the other hand, Homer's Alexandrian editors were much given to removing repeated lines from Homer's text and a line like this appears else-

---

1. Strabo, op. cit., x, 2.12, (cf. I, 2.20, I, 2.28).
2. *Odyssey*, x, 467–471.

where.[1] My own view is that Circe should be located at the southern tropic as Crates of Mallus, on Strabo's account, suggests. In Queensland, Australia, she would be called a 'Rockhampton girl'.

According to Crates of Mallus, Odysseus' journey to the dead took him from the winter tropic toward the South Pole. Let us suppose for a moment that Crates' *Odyssey* lacked the line here about the lengthening days and that Crates held views like Strabo's about Aeaea's location in the southern hemisphere. Let us suppose further that he believed Odysseus to have travelled to the dead in midwinter and that he left Aeaea in that season. Then the heat which drove the stag to water one year earlier could only have occurred where it is still hot in the midwinter, and that is at the tropic. On this view, then, we have some grounds for accepting the surprising exactitude of Crates' location for Aeaea at the winter tropic.

Phaeacia is much easier and in any case we have a clear identification of its location exoceanically by a geographer contemporary with Strabo. The critical passage in Homer's account of Phaeacia comes in his description of the garden of King Alcinous:

> Outside the courtyard but stretching close up to the gates, and with a hedge running down on either side, lies a large orchard of four acres, where trees hang their greenery on high, the pear and the pomegranate, the apple with its glossy burden, the sweet fig and the luxuriant olive. Their fruit never fails nor runs short winter and summer alike. It comes at all seasons of the year, and there is never a time when the West Wind's breath is not assisting, here the bud, and here the ripening fruit; so that pear after pear, apple after apple, cluster on cluster of grapes, and fig upon fig are always coming to perfection. In the same enclosure there is a fruitful vineyard, in one part of which is a warm patch of level ground, where some of the grapes are drying in the sun, while others are gathered or being trodden, and on the foremost rows hang unripe bunches that have just cast their blossom or show the first faint tinge of purple. Vegetable beds of various kinds are neatly laid out beyond the farthest row and make a smiling patch of never failing green. The garden is served by two springs, one led in rills to all parts of the enclosure, while its fellow opposite, after providing a watering-place for the townsfolk, runs under the courtyard

---

1. *Odyssey,* XIX, 153; XXIV, 143.

gate towards the great house itself. Such were the beauties with which the gods had adorned Alcinous' home.[1]

This remarkable garden is mentioned in a strange story told by Diodorus Siculus who lived like Strabo in the first century BC. Diodorus describes an island far out to sea:

> Their climate is most temperate, we are told, considering that they live at the equator, and they suffer neither from heat nor from cold. Moreover, the fruits in their island ripen throughout the entire year, even as the poet writes: 'pear after pear, apple after apple, cluster on cluster of grapes, and fig upon fig, are always coming to perfection.' And with them the day is always the same length as the night, and at midday no shadow is cast of any object because the sun is in the zenith.[2]

This places Phaeacia at the equator, more or less midway between Laestrygonia and Cimmeria on the basis of our analysis so far.

The equatorial location of Phaeacia agrees with the opinion of Eratosthenes and Polybius, as recorded by Strabo, that there was a temperate strip of earth right on the equator.[3] According to Polybius this area is subject to rains because the north winds strike against the mountains in that region. Perhaps unconsciously, these remarks of Polybius bear upon the doom of Phaeacia, which was to be ringed by mountains at the command of Poseidon. Odysseus arrived in Phaeacia in the wake of the most terrible storm. But it is impossible to tell whether Eratosthenes or Polybius derived their belief in the existence of a temperate area right on the equator from reading these lines of Homer in the way that Diodorus did. Strabo upbraids Polybius for failing to recognize that Phaeacia was placed in fiction in the Atlantic Ocean.[4] This leaves open the question of whether either Strabo or Polybius believed Phaeacia to be equatorial.

The simultaneous coexistence of all the seasons or all the stages of growing in the gardens of Alcinous is characteristic of temperate

1. *Odyssey*, VII, 112–132.
2. Diodorus Siculus, *History*, tr. C.H. Oldfather (London: Loeb Classical Library, 1939), II, 56–57.
3. Strabo, op. cit., II, 3.2-3.
4. Ibid., I, 2.18.

locations on the equator. It is also a form of omnipresence which the Phaeacians enjoy. The other form of it is their magic ships. Alcinous describes them to Odysseus:

> I now appoint a day for your conveyance home: tomorrow, let us say. you shall lie there lapped in sleep, while they row you over tranquil seas, till you come to your own country and your house or anywhere else where you would like to go. Nor does it matter if the spot is even more remote than Euboea, which is said to be at the world's end by those of our sailors who saw it, that time they took red-haired Rhadamanthus to visit Tityos, the son of Earth. They not only got there, I must tell you, but finished the return trip also in one and the same day without fatigue. But you shall learn from your own experience the surpassing excellence of my ships and how good my young men are at churning up the sea water with their oars.[1]

And

> you must also tell me where you come from, to what state and to what city you belong, so that my ships as they convey you there may plan the right course in their minds. For the Phaeacians have no steersmen, nor steering-oars such as other craft possess. Our ships know by instinct what their crews are thinking and propose to do. They know every city, every fertile land, and hidden in mist and cloud they make their swift passage over the sea's immensities with no fear of damage and no thought of wreck. [2]

And Athene describes their ships to Odysseus:

> and their ships are as swift as a bird on the wing or as a thought.[3]

It is tempting to see these ships, like the garden, as astrogeographical and exoceanic. On the equator the stars go round faster than they go round near the Poles. On the equator they go fastest of all, since they must cover the widest orbit in the same time. And, of course, they still take exactly the one day to make their orbit, as do all the other stars. If a Phaeacian ship were to follow a star right round the equator, keeping pace with it on the surface of the sea, it would reach every point on that largest circle and from it return in

1. *Odyssey,* vii, 317–328.
2. Ibid., viii, 555–563.
3. Ibid., vii, 36.

the one day. Its return would itself mark the limits of the day which it had taken. Of course, the Phaeacian ships go everywhere on earth, not just round the equator, but since this is the largest circle, it may be taken *a fortiori* to include all other return journeys. These ships have no pilots, helmsmen or steering oars but direct themselves in accordance with their perfect understanding of the thoughts and minds of men. That is why Alcinous needs to know from Odysseus where he is going to, so that he may tell the ship. In this the Phaeacian ships are comparable to the automatic forge described by Homer in the *Iliad* which varies the force of the blast of its twenty bellows in accordance with Hephaestus' spoken commands.[1]

Homer also tells us that the Phaeacians used to live near the Cyclopes in Hyperia but moved to Scherie because the Cyclopes were bad neighbors and stronger than them.[2] When Odysseus recites the tale of his adventures to Alcinous and his court, the third episode which he recounts is his meeting with the Cyclops Polyphemus. Odysseus narrates a journey which begins with the Cyclopes and finishes with his arrival in Phaeacia, in some ways the same journey as the Phaeacians had taken when they moved away from the Cyclopes. This may explain Alcinous' confidence that Odysseus has told them the truth in his recital.[3] In his account of his adventures Odysseus retells to the Phaeacians their own history as Demodocus, their minstrel, retells him his. A long way from the equator is at or near one of the Poles, and it is here perhaps that we should place the Cyclopes, along with the Laestrygonians with whom they share many features. I think the North Pole more likely than the South Pole since Homer seems to dramatize the transition to the southern hemisphere when Odysseus and his companions discuss the location of Aeaea.

According to Strabo, Homer fictionally places Phaeacia in the Atlantic Ocean along with Calypso's island of Ogygia and Odysseus' journey to and from the dead. Alcinous tells Odysseus that the furthest a Phaeacian boat had ever gone was to Euboea in Greece, at

---

1. *Iliad*, XVIII, 469–473.
2. *Odyssey,* VI, 4–6.
3. Ibid., XI, 362–367.

the eastern end of the Mediterranean. This certainly suggests that this boat had to traverse the Mediterranean basin from west to east on its journey from Phaeacia, reaching Euboea from the direction of the Pillars of Heracles or Straits of Gibraltar. But on this assumption, it is very hard to determine from which direction their ship arrived at the Straits of Gibraltar. If Phaeacia was at the equator then their ship would have come from the south, rather than from the west. We do know, however, that wherever Phaeacia was, Ogygia and Calypso were seventeen days' good sailing by raft to the west of it, since this is the length of time Odysseus was on his raft for that journey, driven by the wind Calypso had sent him, keeping the Great Bear on his left and Orion on his right. This same west wind which blew to Phaeacia is mentioned in the account of Alcinous' garden where it is always blowing to make the plants grow and ripen. Of course, if this west wind was always blowing on Phaeacia, then Calypso hardly needed to send it to help Odysseus on his way there. Calypso herself lays claim elsewhere to credit which is not her due, when she tells Odysseus that she has decided to help him on his way out of pity for him, rather than because she was compelled to it by Hermes and the gods.

* * *

Where were Phaeacia and Ogygia, the Cimmerians, Aeaea and Laestrygonia? To these questions we may now give clear answers concerning Phaeacia, the Cimmerians and Laestrygonia, and rather less clear answers concerning Ogygia and Aeaea. This is some small success but we are as yet no nearer to discovering the locations of the Lotus Eaters, Aeolia, the Sirens, Scylla and Charybdis or Thrinacie. From our present point of view, the astrogeographical method of analyzing Homer's account of Odysseus' adventures has only limited application. In fact, as we shall see, the Sirens, Scylla and Charybdis and Thrinacie are all of them open to this kind of interpretation on another plane. As for Aeolia, it was a floating island which must make it particularly hard to locate on any analysis. The Lotus Eaters belong rather with the Cicones who precede them than with the Cyclopes who follow them, and I am not sure that they are exoceanic at all. Like the Cicones they are unexceptionably human except for their strange fruit, while from the encounter with the Cyclopes until the return to Ithaca, all of Odysseus' encounters are uniformly fantastical in one way or another.

According to Strabo, Eratosthenes remarked that the scenes of Odysseus' adventures would be found only when the cobbler was found who sewed up the bag of winds.[1] This, no doubt, was a reply to those who argued with each other over the exact location of the adventures within the Mediterranean basin, or at least within the known world. We have not done this here. Instead we have attempted to explain certain lines in Homer's account as elements in a systematic geography on the largest scale. This way of reading Homer, in theory if not in detail, was developed from at least the second century before Christ by writers such as Crates of Mallus. But interpreters such as Crates were as bitterly attacked by ancient critics as those who located the adventures in particular parts of the known world. According to Strabo, those who read Homer scientifically were regarded as madmen, as though, to use Strabo's words, Homer were no more to be reckoned with than a ditch digger or casual labourer.[2] The way of interpreting Homer's geography which I have followed certainly did not sweep the field of all other interpreters when it appeared, but became one more way of reading Homer along with the others. Nor do we have from the ancient geographers who interpreted Homer scientifically a single coherent account such as the one I have given. Though the interpretations I have offered are to be found in ancient writers, they are widely scattered and are by no means the only scientific interpretations of these passages. Strabo, for example, seems to argue in some passages that Hades is not to the south but to the west, though he also records Crates' view that Odysseus' journey to the dead was from the winter tropic towards the South Pole. Strabo also said that Homer 'had gone to the uttermost bounds of the inhabited world, encompassing the whole of it in his descriptions.'[3] What I

1. Strabo, op. cit., I, 2.15.
2. Ibid., III, 4.
3. Strabo, op. cit., I, 1.2.

have offered above is a brief but systematic reading of Homer's lines as I have understood them, and my reasons for understanding them so.

Then there is the difficult question of how much geography, especially systematic geography of the kind I have discussed, Homer himself could have known. This seems to have been the basis of the ancient attack made on scientific interpreters such as Crates of Mallus if we are to judge from Strabo's sarcastic comparing of Homer to a ditch digger or casual labourer. Strabo defends the scientific interpreters from the charge of madness by arguing that Homer was, indeed, a very wise man and not just anyone. Elsewhere Strabo makes it clear that he supposes Homer a very much better geographer than many of the poets who followed him. Strabo himself would have had little difficulty in accepting that Homer knew what the scientific interpreters claimed he knew, since he believed that Homer knew that the earth was round. It must be said, however, that Strabo's ground for believing this is not very plausible in itself. Strabo argues from the passage in which Odysseus sees the coast of Phaeacia from the top of a big wave, that Homer knew of the curvature of the earth.[1] Whatever the truth of this, Homer's use of astrogeographical phenomena as a means of describing faraway places goes a long way in itself to justify Strabo's claim that he founded the science of geography. In his view Homer adumbrated many methods of systematic geographical analysis which all the geographers who followed him were to use.

But even this formulation hardly does justice to the quotations from the *Odyssey* which I have juxtaposed above. What is remarkable about these quotations is not merely that they are suggestive of certain methods of geographical analysis but that, in the same respect that they are suggestive, they are also integral and complete. We may set out our more assured results in the following table:

---

1. Ibid., I, 1.20.

ZONE	STAR	PEOPLE
Arctic Circle	Great Bear	Laestrygonians Cyclopes?
Tropic of Cancer		
Equator	Orion	Phaeacians
Tropic of Capricorn		Circe?
Antarctic Circle		Cimmerians

I have already suggested that Odysseus' journey is not any journey but that it takes in the limits of latitude, so that in this respect it becomes an archetypal or exemplary journey. Whatever Homer's views may have been about the stereometry of the surface of the earth, his account of Odysseus' travels is, from the point of view of our own knowledge, at once exact and complete. It is easy to imagine how Homer's powerful images of the earth's extremes could have played a part in the intellectual direction of the geographers who followed him, since they, with everyone else, were taught to revere Homer as the major exponent of their tradition. Here, as in many other fields, all roads led back to Homer.

But Homer's geography is also poetic, unlike that of the geographers who followed him. He helps us to feel the physical phenomena of the Poles and the equator from within. There is a connection between the middle position of the equator and the exceptional equanimity and harmony of the Phaeacians, where the king and queen enjoy exactly equal power, where no human enemies ever come against them, and the gods appear to them without disguise. The Ethiopians were often visited by the gods and were believed to be the founders of religion among the peoples of the earth. This was explained by the fact that they lived closest of all peoples to the planetary powers since they lived on the equator which the planets cross, back and forth, on their paths to and from their tropics. The same may be said of the Phaeacians. But more interesting still is the connection which Crates made between the South Pole and the dead. In this instance a geographical zone of the earth's surface becomes the location for a merely psychic reality, on an entirely different plane from the human world. Here the rivers Pyriphlegethon and Cocytus pour their waters into Acheron, the lake at the bottom of the world. The psychic residues of life in the northern hemisphere are gathered as into a sump or drain. It is strangely moving that just when Homer gives one of his most exact accounts of solstial variations at the Pole he is also, as it appears to us, speculating most extravagantly about the afterlife.

In this matter our own education inclines us to the view that these first steps taken by Homer towards a scientific geography were still wrapped in the gloom of superstition, whatever part they may have played in the development of geography over the next few centuries. It is not until Eratosthenes measured the earth and he and Hipparchus established the grid of latitudes and meridians to determine the precise locations and extents of land masses, that geography had assumed its proper form and had at last shed these irrelevancies. On the other hand Homer's characterizations of the earth's limits enabled the assimilation of these geographical principles to much larger schemes of understanding, which the geographical principles are taken to embody and represent. For Homer, the structure which he discerns in the terrestrial and celestial worlds has implications which go far beyond the exact and limited application of this structure in the sciences of astronomy and geography. The Poles, the equator, the tropics not only determined the surfaces of heaven and earth, they were also symbols or mathematical emblems which realized orders of reality which could not be approached by the science of geography as it came to be understood. These geographical principles should be so elevated, if anything should be, since these limits and terms are the largest and most inclusive of all, the first determinations of the physical cosmos as a whole.

Homer describes poetically the geographical regions which he locates scientifically in the passages which I have quoted. From a certain point of view, this poetry is the most valuable single contribution which Homer makes to our own scientific understanding as opposed to our knowledge of its historical development. It is just here, with Homer's guidance, that we can connect these principles of geography and astronomy with the principles of other sciences. The terms of geometry and astronomy, Poles, equator, tropics, are inter-changeable with the terms of other sciences and orders of existence. The other orders of physical existence are contained within the geo-graphical and astronomical cosmos and may be said to be produced by it. If, then, their structure should manifest the same order as the parent structure which bore them, that is what we should expect.

Such an application of the geographical and astronomical princi-ples is suggested by an obscene and charming story told by the early church against the Olympian religion. Like the Hyperboreans, the events recorded here are not found in Homer. But like the Hyper-boreans these events indicate some popular beliefs about the extremities of the earth which are consistent with the account I have given of Odysseus' travels. We may say that Apollo of the far north and Dionysus as described in this story, are polar partners, twin gods who share the seat on the rounded navel stone at Delphi.

> While Dionysus, born at Nysa and son of Semele, was still among men, he wished to become acquainted with the shades below and to inquire into what went on in Tartarus; but this wish was hindered by some difficulties, because from ignorance of the route, he did not know by what way to go and proceed. One Prosumnus starts up, a base lover of the god, and a fellow too prone to wicked lusts, who promises to point out the gate of Dis, and the approaches to Acheron, if the god will gratify him and suffer him to take from the god the pleasures of a wife. The god, without reluctance, swears to put himself in his power and at his disposal, but only immediately on his return from the lower regions, having obtained his wish and desire. Prosumnus politely tells him the way, and sets him on the very threshold of the lower regions. In the meantime, while Dionysus is inspecting and examining carefully Styx, Cerberus, the Furies and all other things, the informer passed from the number of the living, and was buried according to the manner of men, Dionysus comes up from the lower regions, and learns that his guide is dead. But that he might fulfil his promise, and free himself from the obligation of his oath, he goes to the place of the funeral and cutting down the stron-gest branch of a fig tree he carefully works the wood into the shape of a human penis and fixes it on the grave mound. Then baring his backside, he introduces it and sits down on it. Then, imitating the lust of his violator, he twists his buttocks this way and that, and

thinks to suffer from the stick what he had promised with sincerity long before.[1]

It would be hard to find in the records even of Greek and Roman mythology a story more ludicrous and more obscene than this. It deserves a place alongside Hesiod's story of the birth of Aphrodite from the severed genitalia of Uranus.[2] In this form, this story of Dionysus comes to us from Arnobius' treatise *Against the Heathen*. In the English translation of Arnobius before me much of the story is given in Latin. The treatise was written about 200 CE. Arnobius told the story to explain the mysteries, as he called them, in which Greece erected phalli in honour of father Bacchus, and the whole country was covered with images of men's genitals. It was clearly only with considerable diffidence that Arnobius could bring himself to mention and explain such things, so ashamed was he. The story he tells about Dionysus and Prosumnus shows that the infamous display of the phalli was surpassed by an even more infamous myth. 'Are these the gods you force us to worship?' Arnobius asks.

Leaving aside the justice of Arnobius' argument, we may examine this story of Dionysus from an entirely different angle. The story turns on the connection between Dionysus' being shown the entrance to the lower regions and his being sodomized by Prosumnus. In the story these two deeds are made equivalent by the oath which Dionysus upholds with such punctilio. Showing Dionysus the way to Hades is requited by sodomizing him. This may be because the anus is the lowest limit of the human trunk as Tartarus is the lowest limit of the earth. If we take it that the human microcosm is an image of the whole earth in miniature, then the Poles of the human frame are the limits of the spine. The North Pole corresponds to the skull and especially the foramen or fontanelle while the South Pole corresponds to the sacrum. Prosumnus shows Dionysus the way into Tartarus at the bottom of the earth and requires that Dionysus do exactly the same for him in return. Each

---

1. *Ante Nicene Fathers*, vol. 7, eds. Roberts and Donaldson (Michigan: Eerdmans, Arnobius, 1981), 'Against the Heathen', v, 28.

2. *Hesiod, The Homeric Hymns and Homerica*, tr. H.G. Evelyn-White (London: Loeb Classical Library, 1914), 'Theogony', 154–206.

of us, we might say, has his or her own Tartarus. But this assimilation of the human frame to the geographical and astronomical order is an illustration of the way in which the scientific principles of geography and those of anatomy are comparable.

Homer does not mention the Hyperboreans by name in the *Iliad* nor in the *Odyssey*. But Herodotus details some of the ways in which the Greeks conceived of them.[1] At some earlier time the people of the far north had sent a mission of two girls, with an escort of five men, to the Greek island of Delos which was sacred as the birthplace of Apollo. The escort was highly honoured at Delos where they were called *Perpherees* which means literally 'those who carry round', but neither they nor the maidens ever returned to their home. The tomb of the girls was on Delos. There the young men and maidens of Delos cut their hair in honour of the girls from the far north. The boys wound their hair round green stalks and the girls wound theirs round spindles, and they laid these offerings on the tomb. In another ceremony for two other girls from the far north buried on Delos, the Delians collected the ashes of the thigh bones burnt in sacrifice on the altar and scattered them on the tomb of the girls. These Delian rituals realized the closeness of their visitors to the North Pole and to the Great Bear which is also the thigh. Is it too fanciful to see in those locks of hair, wrapped round their sticks, the spiralling motions of the sun, moon and planets as they are carried round above the equator, as seen from the crown of the world where their entire revolutions are visible?

1. Herodotus, op. cit., IV, 33–35.

2

HOMER'S
SPACE ODYSSEY

THIS CHAPTER CONSIDERS the penultimate stage of Odysseus'
return to Ithaca, his journey from Circe to Calypso, from Aeaea to
Ogygia. On this journey he encountered the Sirens, Scylla and
Charybdis, and Thrinacie. We are given several accounts of each of
these adventures, by Teiresias and Circe as well as by Odysseus. As
with the earlier adventures, ancient and modern commentators have
identified the various sites of this penultimate journey all over the
Mediterranean basin and beyond. Charybdis, the whirlpool, has
attracted the most attention. But these adventures, too, are suscepti-
ble of an astrogeographical or exoceanic interpretation, which complements and extends the interpretation given of the earlier
adventures in the first chapter. There we saw how those adventures,
particularly with the Laestrygonians, Phaeacians, and Cimmerians,
constituted a kind of latitudinal analysis of the earth's surface. This
interpretation gave Homer some claim to be regarded as the founder
of scientific geography in Greece and we supported the interpretation with quotations from a range of earlier geographers including
Strabo, Crates, Polybius, and Diodorus Siculus. In this chapter we
shall see how his penultimate journey supplies a complementary
and closely integrated pattern of metaphors for the planets.

Whether or not Homer supposed that the earth was round, the
earlier adventures may be taken to provide an account of the three
limits of the earth's surface; the two Poles and the equator. Homer
defines these limits by means of seasons and especially solstitial
variations, and perhaps by the visibility or otherwise of some of the
fixed stars. These seasonal and solstitial variations are occasioned by

the movement of the sun north and south of the equator, a regular movement which never takes the sun more than 24° either north or south of that line. Likewise, the moon, Mercury, Venus, Mars, Jupiter and Saturn move in their cycles north or south of the equator but never exceed limits close to those which constrain the movements of the sun. From our point of view on earth, the movements of the sun and moon are strictly comparable with those of the planets which I have mentioned, and from this geocentric point of view their movements may all be considered together. To the description of the earth's surface which we have derived from Homer's account of the earlier adventures, Odysseus' penultimate journey adds an account of these heavenly movements viewed geocentrically. From this point of view, the earth, whatever its shape, lies on the axis around which turn the seven stars of the solar system, extending outwards from the earth as far as Saturn, the furthest away. These seven stars all move around the earth more or less within the limits set by the sun's course north and south. In this way they form a kind of disk with the earth at the center and with the motions of the planets, including the sun and moon, forming concentric rings in the same plane. This system is, I suggest, the origin of the whirlpool Charybdis which is a symbol of the solar system viewed geocentrically. The whirling planetary plane is represented by the circling waters of the whirlpool while its center is the axis on which the earth rests. Similarly the sheer cliff of Scylla is the planetary plane, her cave halfway up is the axis and her six heads the six planetary powers other than the sun. The Sirens sing the music of the spheres and the seven herds of fifty cattle and the seven flocks of fifty sheep on the island of Thrinacie represent the days and nights of the year, divided into weeks and rounded off. Here then, in brief, is a way of reading the penultimate journey which complements and extends the astrogeography of the earlier adventures.

Each of these propositions about Odysseus' later adventures will be discussed in this chapter, as well as the meaning of his journey past the planets. But I must first make clear that the interpretation about to be given is not original with me. In the first chapter I derived the account which I gave of the earlier adventures from a selection of comments made by early geographers about Odysseus' wanderings. In this chapter I have derived my interpretation from a reading of a passage in Plato, who lived long before those geographers, though long after Homer. At the end of the *Republic* Socrates tells the story of Er who visits the underworld. At the very beginning Plato explicitly contrasts this story with the account that Odysseus gives to King Alcinous. But in Plato's story many of the characters from Homer's underworld reappear and so also do the Sirens. I do not know how far Plato intended his myth to elucidate Homer or whether he intended it to replace Homer. But I do know that without Plato's myth I could not have arrived at the interpretation of Homer given in this chapter.

Here is Socrates with the *Myth of Er*:

> Well, I said, I will tell you a tale; not one of the tales which Odysseus tells to the hero Alcinous, yet this too is a tale of a hero, Er the son of Armenius, a Pamphylian by birth. He was slain in battle, and ten days afterwards, when the bodies of the dead were taken up already in a state of corruption, his body was found unaffected by decay, and carried away home to be buried. And on the twelfth day, as he was lying on the funeral pile, he returned to life and told them what he had seen in the other world. He said that when his soul left the body he went on a journey with a great company, and that they came to a mysterious place at which there were two openings in the earth; they were near together, and over against them were two other openings in the heaven above. In the intermediate space there were judges seated, who commanded the just, after they had given judgement on them and had bound their sentences in front of them, to ascend by the heavenly way on the right hand; and in like manner the unjust were bidden by them to descend by the lower way on the left hand; these also bore the symbols of their deeds, but fastened on their backs. Er drew near, and they told him that he was to be the messenger who would carry the report of the other world to men, and they bade him hear and see all that was to be heard and seen in that place. Then he beheld and saw on one side the souls departing at either opening of heaven and earth when sentence had been given on them; and at the two other openings other souls, some ascending out of the earth dusty and worn with travel, some descending out of heaven clean and bright. And arriving ever and anon they seemed to have come from a long journey, and they went forth with gladness into the meadow, where they encamped as at a festival; and those who knew

one another embraced and conversed, the souls which came from earth curiously enquiring about the things above, and the souls which came from heaven about the things beneath. And they told one another of what had happened by the way, those from below weeping and sorrowing at the remembrance of the things which they had endured and seen in their journey beneath the earth (now the journey lasted a thousand years), while those from above were describing heavenly delights and visions of inconceivable beauty.

The story, Glaucon, would take too long to tell; but the sum was this:—He said that for every wrong which they had done to any one they suffered tenfold; or once in a hundred years—such being reckoned to be the length of man's life, and the penalty being thus paid ten times in a thousand years. If, for example, there were any who had been the cause of many deaths, or had betrayed or enslaved cities or armies, or been guilty of any other evil behavior, for each and all of their offences they received punishment ten times over, and the rewards of beneficence and justice and holiness were in the same proportion. I need hardly repeat what he said concerning young children dying almost as soon as they were born. Of piety and impiety to gods and parents, and of murderers, there were retributions other and greater far which he described. He mentioned that he was present when one of the spirits asked another, 'Where is Ardiaeus the Great?' (Now this Ardiaeus lived a thousand years before the time of Er: He had been the tyrant of some city of Pamphylia, and had murdered his aged father and his elder brother, and was said to have committed many other abominable crimes.) The answer of the other spirit was: 'He comes not hither and will never come.' 'And this,' said he, 'was one of the dreadful sights which we ourselves witnessed. We were at the mouth of the cavern, and, having completed all our experiences, were about to re-ascend, when of a sudden Ardiaeus appeared and several others, most of whom were tyrants; and there were also besides the tyrants private individuals who had been great criminals: they were just, as they fancied, about to return into the upper world, but the mouth, instead of admitting them, gave a roar, whenever any of these incurable sinners or some one who had not been sufficiently punished tried to ascend; and then wild men of fiery aspect, who were standing by and heard the sound, seized and carried them off; and Ardiaeus and others they bound head and foot and hand, and threw them down and flayed them with scourges, and dragged them along the road at the side, carding them on thorns like wool, and declaring to the passers-by what were their crimes, and that they were being taken away to be cast into hell. And of all the many terrors which they had endured, he said that there was none like the terror which each of them felt at that moment, lest they should hear the voice; and when there was silence, one by one they ascended with exceeding joy. These, said Er, were the penalties and retributions, and there were blessings as great.

Now when the spirits which were in the meadow had tarried seven days, on the eighth they were obliged to proceed on their journey, and, on the fourth day after, he said that they came to a place where they could see from above a line of light, straight as a column, extending right through the whole heaven and through the earth, in color resembling the rainbow, only brighter and purer; another day's journey brought them to the place, and there, in the midst of the light, they saw the ends of the chains of heaven let down from above: for this light is the belt of heaven, and holds together the circle of the universe, like the under-girders of a trireme. From these ends is extended the spindle of Necessity, on which all the revolutions turn. The shaft and hook of this spindle are made of steel, and the whorl is made partly of steel and also partly of other materials. Now the whorl is in form like the whorl used on earth; and the description of it implied that there is one large hollow whorl which is quite scooped out, and into this is fitted another lesser one, and another, and another, and four others, making eight in all, like vessels which fit into one another; the whorls show their edges on the upper side, and on their lower side all together form one continuous whorl. This is pierced by the spindle, which is driven home through the center of the eighth. The first and outermost whorl has the rim broadest, and the seven inner whorls are narrower, in the following proportions -- the sixth is next to the first in size, the fourth next to the sixth; then comes the eighth; the seventh is fifth, the fifth is sixth, the third is seventh, last and eighth comes the second. The largest (of fixed stars) is spangled, and the seventh (or sun) is brightest; the eighth (or moon) colored by the reflected light of the seventh; the second and fifth (Saturn and Mercury) are in color like one another, and yellower than the preceding; the third (Venus) has the whitest light; the fourth (Mars) is reddish; the sixth (Jupiter) is in whiteness second. Now the whole spindle has the same motion; but, as the whole revolves in one direction, the seven inner circles move slowly in the other, and of these the swiftest is the eighth; next in swiftness are the seventh, sixth, and fifth, which move together; third in swiftness appeared to move according to the law of this reversed motion the fourth; the third appeared fourth and the second fifth.

The spindle turns on the knees of Necessity; and on the upper surface of each circle is a siren, who goes round with them, hymning a single tone or note. The eight together form one harmony; and round about, at equal intervals, there is another band, three in number, each sitting upon her throne: these are the Fates, daughters of Necessity, who are clothed in white robes and have chaplets upon their heads, Lachesis and Clotho and Atropos, who accompany with their voices the harmony of the sirens - Lachesis singing of the past, Clotho of the present, Atropos of the future; Clotho from time to time assisting with a touch of her right hand the revolution of the outer circle of the whorl or spindle, and Atropos with her left hand touching and guiding the inner ones, and Lachesis laying hold of either in turn, first with one hand and then with the other.

When Er and the spirits arrived, their duty was to go at once to Lachesis; but first of all there came a prophet who arranged them in order; then he took from the knees of Lachesis lots and samples of lives, and having mounted a high pulpit, spoke as follows: 'Hear the word of Lachesis, the daughter of Necessity. Mortal souls, behold a new cycle of life and mortality. Your genius will not be allotted to you, but you choose your genius; and let him who draws the first lot have the first choice, and the life which he chooses shall be his destiny. Virtue is free, and as a man honours or dishonors her he will have more or less of her; the responsibility is with the chooser—God is justified.' When the Interpreter had thus spoken he scattered lots indifferently among them all, and each of them took up the lot which fell near him, all but Er himself (he was not allowed), and each as he took his lot perceived the number which he had obtained. Then the Interpreter placed on the ground before them the samples of lives; and there were many more lives than the souls present, and they were of all sorts. There were lives of every animal and of man in every condition. And there were tyrannies among them, some lasting out the tyrant's life, others which broke off in the middle and came to an end in poverty and exile and beggary; and there were lives of famous men, some who were famous for their form and beauty as well as for their strength and success in games, or, again, for their birth and the qualities of their ancestors; and some who were the reverse of famous for the opposite qualities. And of women likewise; there was not, however, any definite character among them, because the soul, when choosing a new life, must of necessity become different. But there was every other quality, and they all mingled with one another, and also with elements of wealth and poverty, and disease and health; and there were mean states also. . . .

And according to the report of the messenger from the other world this was what the prophet said at the time: 'Even for the last comer, if he chooses wisely and will live diligently, there is appointed a happy and not undesirable existence. Let not him who chooses first be careless, and let not the last despair. And when he had spoken, he who had the first choice came forward and in a moment chose the greatest tyranny; his mind having been darkened by folly and sensuality, he had not thought out the whole matter before he chose, and did not at first sight perceive that he was fated, among other evils, to devour his own children. But when he had time to reflect, and saw what was in the lot, he began to beat his breast and lament over his choice, forgetting the proclamation of the prophet; for, instead of throwing the blame of his misfortune on himself, he accused chance and the gods, and everything rather than himself. Now he was one of those who came from heaven, and in a former life had dwelt in a well-ordered State, but his virtue was a matter of habit only, and he had no philosophy. And it was true of others who were similarly overtaken, that the greater number of them came from heaven and therefore they had never been schooled by trial, whereas the pilgrims who came from earth, having themselves suffered and seen others suffer, were not in a hurry to choose. And owing to this inexperience of theirs, and also because the lot was a chance, many of the souls exchanged a good destiny for an evil or an evil for a good. For if a man had always on his arrival in this world dedicated himself from the first to sound philosophy, and had been moderately fortunate in the number of the lot, he might, as the messenger reported, be happy here, and also his journey to another life and return to this, instead of being rough and underground, would be smooth and heavenly.

Most curious, he said, was the spectacle—sad and laughable and strange; for the choice of the souls was in most cases based on their experience of a previous life. There he saw the soul which had once been Orpheus choosing the life of a swan out of enmity to the race of women, hating to be born of a woman because they had been his murderers; he beheld also the soul of Thamyras choosing the life of a nightingale; birds, on the other hand, like the swan and other musicians, wanting to be men. The soul which obtained the twentieth lot

chose the life of a lion, and this was the soul of Ajax the son of Tela-
mon, who would not be a man, remembering the injustice which was
done him in the judgement about the arms. The next was Agamemnon, who took the life of an eagle, because, like Ajax, he hated
human nature by reason of his sufferings. About the middle came the
lot of Atalanta; she, seeing the great fame of an athlete, was unable to
resist the temptation: and after her there followed the soul of Epeus
the son of Panopeus passing into the nature of a woman cunning in
the arts; and far away among the last who chose, the soul of the jester
Thersites was putting on the form of a monkey. There came also the
soul of Odysseus having yet to make a choice, and his lot happened
to be the last of them all. Now the recollection of former toils had
disenchanted him of ambition, and he went about for a considerable
time in search of the life of a private man who had no cares; he had
some difficulty in finding this, which was lying about and had been
neglected by everybody else; and when he saw it, he said that he
would have done this had his lot been first instead of last, and that he
was delighted to have it. And not only did men pass into animals, but
I must also mention that there were animals tame and wild who
changed into one another and into corresponding human natures—
the good into the gentle and the evil into the savage, in all sorts of
combinations.

All the souls had now chosen their lives, and they went in the order
of their choice to Lachesis, who sent with them the genius whom
they had severally chosen, to be the guardian of their lives and the
fulfiller of the choice: this genius led the souls first to Clotho, and
drew them within the revolution of the spindle impelled by her
hand, thus ratifying the destiny of each; and then, when they were
fastened to this, carried them to Atropos, who spun the threads and
made them irreversible, whence without turning round they passed
beneath the throne of Necessity; and when they had all passed, they
marched on in a scorching heat to the plain of Forgetfulness, which
was a barren waste destitute of trees and verdure; and then towards
evening they encamped by the river of Unmindfulness, whose water
no vessel can hold; of this they were all obliged to drink a certain
quantity, and those who were not saved by wisdom drank more than
was necessary; and each one as he drank forgot all things. Now after
they had gone to rest, about the middle of the night there was a thun-
derstorm and earthquake, and then in an instant they were driven
upwards in all manner of ways to their birth, like stars shooting. He
himself was hindered from drinking the water. But in what manner

or by what means he returned to the body he could not say; only, in the morning, awaking suddenly, he found himself lying on the pyre.[1]

This long narrative about the afterlife is worth a full-length study of its own, but here we are concerned only with its relationship to the account of Odysseus' adventures in the *Odyssey*. It is striking that Plato's narrative begins with a denial that it is one of Odysseus' stories to King Alcinous but ends with an account of Odysseus' choosing his next life. Moreover, we meet again in Plato's narrative some of the souls whom Odysseus met when he visited the dead in the *Odyssey*. The single most obvious difference between these accounts of Homer and Plato is that in Homer's story there is no clear mention of such dead souls ever being reborn to another mortal life, while this very process is the major topic in Plato's account. But even here Plato may not be so far removed from the spirit of Homer's story. For in Homer's story the souls whom Odysseus meets are all of them drawn to the blood in the ditch which he has prepared, and they are anxious to drink it. This makes for one of the most disgusting and pitiful scenes in the *Odyssey* as the souls of the mighty dead reach down to drink the blood of slaughtered sheep from a hole dug in the ground. These souls, it seems, are eager to grasp even this chance of surrogate life, so desperate is their condition in Hades. Some ancient commentators supposed that the milk and honey used in Odysseus' conjuration of the dead in the *Odyssey* have this same power to draw these dead souls, because they are the food given to infants.[2] If, then, Homer's souls of the dead are as anxious as this to regain a mortal existence on earth, Plato's account of their actually being reborn realizes their dearest wish as Homer describes them.

The single greatest similarity between Plato's and Homer's accounts is between the spindle and whorl on the lap of Necessity in Plato's account and the adventures with the Sirens, Scylla and Charybdis in the *Odyssey*. Plato's account of the Sirens in the *Myth of*

1. Plato, *The Republic*, tr. B. Jowett, in *The Dialogues of Plato* (London: Encyclopaedia Brittanica, 1952), 614B–621B.

2. *Thomas Taylor the Platonist*, op. cit., Porphyry, 'Concerning the Cave of the Nymphs', p316.

Er suggested to me that the adventure of the Sirens in the *Odyssey* might represent one aspect of a cosmic system similar to the one which Plato describes with his spindle and whorl. From there I conjectured that a single complex system such as Plato makes of the heavens in the *Myth of Er* might be represented in the *Odyssey* by a series of discrete adventures which are in effect all the same adventure. In this way I came to read Scylla and Charybdis, with the Sirens, as a series of partial metaphors for the planetary system. Plato's narrative provided the clue to understanding Homer at this point, which is the more surprising because Plato begins his narrative by emphasizing its difference from Odysseus' stories to Alcinous.

But even in respect of this homology which I discerned between the narratives of Homer and Plato, there was a major difference in their accounts. In the *Myth of Er* the souls pass through the planets on their way into another mortal life. They freely choose what their new life will be, and their choice is ratified by their passing before each of the three Fates who are enthroned among the planets. In the case of Homer's Odysseus, however, the reverse is true. Odysseus, as I understand it, is striving to escape the planetary powers which seek to entrap him. Odysseus is escaping past the world of the planets and the reward for his succeeding is the immortality offered by Calypso, though he does not take up her offer. On the other hand the direct and inevitable consequence of re-entering the planetary realm as a mortal creature is death at the end of it, so that Homer may fairly represent the lot of those who are seized by the planetary powers as death. From this point of view we may say that the heaps of corpses which surround Homer's Sirens on their island correspond to the souls whom Plato describes on their way to another mortal life, as do the six companions of Odysseus whom Scylla snatches up to her cave which leads down to darkness and Erebus. From Plato's story we may infer that each one of us once ratified our choice of life with the Fates, though we can no longer remember it since we drank the waters of Unmindfulness. And the price of that choice, whatever else it was, was death. To read it this way, Odysseus reneged on his choice which placed him beneath the planets in a mortal existence as Odysseus. To escape the consequences of his having made the choice, he had to row beyond the planets' reach and this is the story of his penultimate journey in the *Odyssey*. His success in this enterprise is unique; he is the sole survivor. His reward is the offer of immortality. In this respect Plato's *Myth of Er* is the tally to Homer's account of the later adventures. It is true that Odysseus rows past the Sirens, Scylla and Charybdis, not from them as he would if he were exactly to reverse the journey of Plato's souls on their way to the earth at the center. But Odysseus resists that centripetal pressure and does not succumb to it. It is in this that he is the opposite of Plato's souls.

Which element of the Sirens' song is more seductive, the music or the words? Their singing is sweet as honey musically but the words reveal all that happens on the earth. They know it is Odysseus who is coming towards them for they call him by name and promise to tell him everything that happened during the war at Troy. The power of the Sirens consists in this combination of beautiful music and omniscience. It appears from Homer's account that they sing in unison and that there are two of them, for Homer always speaks of them in the dual number. In Plato's account there are eight Sirens, each of whom emits a single tone of the octave as she is carried round by the revolution of the ring on which she stands. The words of Homer's Sirens Plato allots to the three Fates enthroned among the rings who sing in harmony with the notes of the Sirens, Lachesis of the past, Clotho of the present and Atropos of the future. Leaving aside for a moment the question of the number of the Sirens, we can see that Plato's distribution of the song of Homer's Sirens between his Sirens and the Fates makes clearer the power in that combination of music and words which is a feature of Homer's adventure. For the song of Plato's three Fates is in tune with the sounds made by his Sirens. It is, I think, Plato's view that while the movements of the heavens do not cause or influence the destinies freely chosen by mortal creatures, they do mark these destinies. The free choice of a life from the assortment offered is ratified by the journeying of the souls past the Fates and the planets, so that the events of the new life and the dispositions of the heavens are coordinated. Hence the story of what is past or passing or to come is co-ordinated with the movements of the rings on which the Sirens stand. Certainly from the time of Pythagoras, it was believed that the motions of the stars made music. In the story of Er this music marked and expressed events in the world, as did the story sung by the Fates. From this angle, the tones of the Sirens and the words of the Fates signify exactly the same thing; they are equal and parallel expressions of these destinies in the world. And this in turn illuminates the relation between music and words in Homer's story of the Sirens.

There are two major differences between Homer's Sirens and those of Plato: number and motion. There are two Sirens according to Homer, compared to Plato's eight, and Homer's Sirens are static on their island while Plato's Sirens revolve. Taken by themselves, it would be hard to see how Homer's and Plato's Sirens could be the same, but we must remember that the Sirens may be only one of the Homeric adventures which constitute Homer's planetary structure. Scylla and Charybdis each contribute other elements to the complex. In Plato's account, on the other hand, these three adventures are combined in a single narrative. If, then, there are elements of Plato's account of the Sirens which are not found in or which contradict Homer's account of the Sirens, this does not by itself show that their Sirens are not the same. Provided only that the elements in Plato's Sirens which are missing from Homer's Sirens are nonetheless present in Homer's accounts of Scylla and Charybdis, we may still claim that Homer's and Plato's narratives are essentially the same. And this is true of these two differences between their accounts of the Sirens. For in the case of number we approach much closer to Plato's eight Sirens with the six heads of Scylla; and in the case of motion, the whirlpool of Charybdis and the movements of Scylla's heads provide the parallel with Plato's moving Sirens. In Homer's account the Sirens are simplified to the point where they represent merely the seductive musical power of the world, whose history and music combine in their marvellous song. Their other powers and properties are transferred to their reappearances in other forms in other adventures. In order fully to realize the significance of their musical storytelling Plato has in his turn to transfer some of their powers to other elements in his narrative when he leaves the Sirens with their music and gives their storytelling to the three Fates.

I do not know why Homer has two Sirens rather than some other number. On the other hand, I cannot see how multiplying the number of Sirens would contribute much to Homer's story. In the case of the heads of Scylla, their number, independence of each other, and three rows of teeth all add to the terror. If we have to choose which of the rings of Plato's heavens Homer's two Sirens represent, it is easy to nominate the sun and the moon, or to suppose that by his two Sirens Homer is distinguishing between all the orbits of the fixed stars on the one hand and those of the seven planets on the other. Homer's Sirens remain on their island while Odysseus is rowed past them, bound to the mast, and on a sea becalmed. Er, on the other hand, stands watching the revolutions of the Sirens on their rings. When Odysseus returns to Charybdis after the destruction of his ship, he clings for several hours to a fig tree standing above the whirlpool. This second encounter with Charybdis relates Odysseus to the whirlpool as Er stands to the revolving planets. His posture there and his being bound to the mast are similar. Holding fast to the mast or tree is his means to salvation from the dangers of the Sirens and Charybdis. In this respect the second encounter with Charybdis has the same form as the adventure with the Sirens, except that the relation between the moving and the unmoving is reversed.

The second adventure of this penultimate journey has remained, like the Lotus-Eaters, part of the common lore. The choice which Odysseus has to make between Scylla and Charybdis still serves as a metaphor for any choice between catastrophic alternatives. Circe tells Odysseus that it is better to lose six men than all of them. This is the less disastrous of his choices, to steer clear of Charybdis and let Scylla have her way with them. In fact, as Circe represents it, Odysseus has a choice to avoid Scylla and Charybdis altogether, both of them, and go another way. But this way leads between two other deadly cliffs, through the Clashing Rocks which have never yet permitted a ship to pass, except the Argo which was saved by Hera because she loved Jason. Not even the doves which carry ambrosia to Zeus pass between these rocks with impunity, but one of them is always lost so that Zeus has to send another to make up the number. This sounds like the unavoidable price of passing beneath Scylla's cave, but the Clashing Rocks claim even the birds. So the impossible choice between Scylla and Charybdis is repeated in the impossible choice between Scylla and Charybdis on the one hand and the Clashing Rocks on the other. An additional disincentive to choosing the Clashing Rocks is that their passage risks storms of fire as well. Otherwise, these Clashing Rocks are very like Scylla and Charybdis, making Odysseus' choice between them a distinction without a difference.

The parallels between the Clashing Rocks and Scylla and Charybdis are obvious: more obvious still are those between Scylla and Charybdis themselves. Each is a divine power who draws sailors into a hole or lair by force. In each case the end is at the center, since, like the spout at the center of the whirlpool, Scylla's cave is in the middle of the cliff. In this respect Scylla's cliff and cave are the whirlpool and spout of Charybdis represented vertically. Circe tells Odysseus that the distance from Scylla's cliff to Charybdis is a bowshot, but the distance from the sea to Scylla's cave is further than a bowshot. Since an arrow shot vertically reaches less far than an arrow shot laterally, these distances which Circe gives suggest that the cave and the spout are more or less equidistant from the surface of the water at the base of Scylla's cliff. Scylla and Charybdis are, as it were, reflections of each other, and they seem to work in a symbiotic manner. Scylla takes advantage of the sailors' desire to escape Charybdis and Charybdis, no doubt, enjoys wrecking those who are terrified by Scylla. Scylla and Charybdis are the same danger represented in two different planes. Homer's description of the passage between them induces vertigo, just as in heaving seas sailors must sometimes walk on bulkheads as though they were decks. Wherever he looks to escape the awful danger of Scylla, Odysseus is confronted by the same danger in another form, like a hall of distorting mirrors.

The sound effects which Homer uses in his account of this second adventure are as memorable as those of the Sirens. So terrible is the noise of Charybdis as the ship approaches that Odysseus' crew let go of their oars in horror and the ship comes to a standstill. These sounds are caused by the reverberating gurgitations and regurgitations of the water by Charybdis. Homer's account of Charybdis at this point is powerfully onomatopoeic and alliterative, making particular use of the clashing consonants at the end of Charybdis' name. This sucking, roaring, whirring sound is matched by Scylla whose voice is like the yelp of a newly born puppy, a litter of them since she has six heads. It is a terrible sound, a *sopranino* accompaniment to the bass tones of Charybdis. By describing the yelping of new born puppies as terrible, Homer converts our sentimental associations with new life into the horror we feel for Scylla's teeth. Everything here conspires to unman Odysseus and his crew, and just as their eyes are drawn irresistibly to the whirlpool, Scylla seizes her chance and takes her six men. The noise must have been enormously magnified by the echoes from the two cliffs. Scylla's cliff, we are told, is always clouded at the top with dark blue clouds and when Charybdis sucks down the sea the sand below is revealed, as dark as the clouds. When Charybdis spews out the water, spray flies as high as the cliff tops so that this enclosed space between the cliffs, the sea and the clouds is as much vapor as air. It is a scene as fantastic and horrible as the one which Circe describes to Odysseus when she directs him to the dead.

The whirlpool is an image of the planetary orbits since both whirlpool and orbits are plane surfaces which turn on their axes at their centers with a spinning motion. In Plato's story the souls to be reborn cross these orbits on their way to the center which is earth. Odysseus, on the contrary, is seeking to escape these orbits and succeeds, first by rowing close to Scylla's cliff and then by clinging to the fig tree. Scylla's cliff is so smooth that even with twenty hands and legs a man could not climb it. The cave is in the middle of the cliff and Scylla is up to her middle in the cave. The upper half of her body and the necks of her six heads are long enough to enable her to take her victims from the sea beneath, longer than a good vertical bowshot. In the cave also, it appears from Circe's description, lives Scylla's mother, the Mighty One. Like the monsters of Beowulf, Scyllas live in families. The cave is also described as a pit or gulf and it goes down 'to darkness and Erebus'. This phrase 'to darkness and Erebus' refers, I think, to the direction of the cave as it leads away from the cliff face, not to the direction which the cave faces at the cliff face, as it is usually translated. Though Homer's word for darkness may sometimes refer to the west or the north as the two directions of darkness in the northern hemisphere, it also just means darkness, especially when linked with another word denoting death as here. To be caught by one of Scylla's heads is to be set on a course for Hades. In any case, in the southern hemisphere where Strabo places Aeaea, Odysseus' last port of call, the darkness is towards the south, not the north.[1]

Scylla has six heads and twelve feet though we are not told whether these twelve feet are inside or outside the cave. Like the twelve axeheads of Odysseus' trial with the bow, these twelve feet may signify twelve zodiacal constellations, the fixed stars through which the planets pass in their orbits. In that case the six heads and the feet would constitute seven of the eight rings which Plato describes in the *Myth of Er.* This is much closer to the Platonic number than the number of Homer's Sirens. The ring omitted in Homer's account of Scylla is likely to be the sun, to which a whole adventure is devoted in the story of the Cattle of the Sun. The six heads are remarkable because they move independently of each other, on the ends of their tentacular necks. There are very few creatures in nature with two heads, so that Scylla's form realizes very well the exceptional diversity in unity of the planets. The three rows of teeth, thick and close, may correspond to the suckings down and regurgitations of Charybdis three times a day, or even to the three phases of a planet's course, at its tropics and the equator.

The Sirens of Homer provide the music, Charybdis the motion and Scylla the numbers; these three adventures, taken together, constitute the cosmological order which Plato describes in the *Myth of Er.* At the same time Plato's synthesis in the *Myth of Er* comprises other elements from Odysseus' travels, his encounters with several dead heroes, and with Circe in her capacity as a transformer of humans into animals and animals into humans. At the end of the *Myth of Er* Plato, too, describes how souls change from one species to another. Of all those whom Er observes, only Odysseus lives his next life as a human being of the same gender as before. The great difference between the accounts of Homer and Plato is not in the scene, nor even the direction of the action, but in the tone. The

---

1. Strabo, op. cit., x, 2.12 (cf. I, 2.20; I, 20.28.).

rebirth of the souls in the *Myth of Er* is conducted with a benign decorum and the vista of the planets which is revealed to them is noble and soothing. In Homer's account these same elements are more often terrible and violent. In Homer's story the gods of heaven are represented in their malefic and demoniacal aspect. In this respect Homer's spirit is closer to the Orphic and Pythagorean traditions than to the *Myth of Er* which follows them. In those earlier traditions, as in the Egyptian, mortal life was often represented as a prison which the planets guarded or as a mire from which we must rise up and be cleansed.

Odysseus' last adventure on this penultimate journey is with the Cattle of the Sun on the island of Thrinacie. He has been advised in some detail on how to conduct himself there by both Teiresias and Circe. I have suggested that the seven herds of fifty cattle and the seven herds of fifty sheep represent the days and nights of the year arranged in weeks and rounded off to the nearest decad. This interpretation places Homer's Cattle of the Sun in the same category as other early accounts of the days of the year. For example, the Greek poet Cleobulus of Lindos, who lived in the sixth century BC and was accounted one of the seven wise men of Greece, propounded the following riddle:

> One sire there is, he has twelve sons, and each of these has twice thirty daughters different in feature; some of the daughters are white, the others again are black; they are immortal and yet they all die.[1]

Who is the father? The answer is the year. Cleobulus counts 360 days and nights, not the 350 of Homer, and his days and nights are girls, not cows and sheep. But both girls and animals are immortal, and they are all animations, as it were, of the days and nights of the year which are thereby given an immediate reality.

The second account comparable to Homer's Cattle of the Sun comes in the story told about Hermes in the Homeric *Hymn to Hermes*.[2] This is also from the sixth century BC. In the story Hermes, as

---

1. *Greek Elegy and Iambus*, tr. J.M. Edmonds (London: Loeb Classical Library, 1914), Cleobulus, I, p159.
2. *Hesiod, The Homeric Hymns and Homerica*, op. cit., 'Homeric Hymn to Hermes', pp363–405.

soon as he is born gets up and with a clever trick steals fifty of Apollo's cattle. Eventually Hermes and Apollo strike a deal over these cattle. Hermes gives Apollo in exchange his tortoise shell lyre which he had made as soon as he got up. If we remember the French word for Wednesday, *mercredi*, we may guess that these fifty cattle which Hermes steals are the fifty Wednesdays of the year. His theft of them explains how they became sacred to him. On this reading we have a much more exact correspondence with Homer's Cattle of the Sun, both in number and kind. Taking these three accounts together we can see how the early Greeks reified what have become for us the abstract notions of the calendar days. According to the Greek understanding these days were realities, equivalent to living creatures on the earth but superior to them in being immortal. Their immortality clearly consisted in their reappearing every year for as long as the sun kept to its course. The Greeks were not alone in thinking that the days were determinate and reverend realities of this kind. The Egyptians revered the thirty-six gods of the Decans, the ten-day periods into which their year was divided, and the Christian calendar likewise animated the days of the year by appointing them to saints.

The Cattle of the Sun are immortal, yet Odysseus' companions kill and cook some of them. The effects are bizarre. The hides of the slaughtered cattle creep along the ground and the meat, raw and cooked, continues to groan. What must have been the effect of eating this immortal beef on the human digestive system? Very little, directly. The companions are on the island another seven days, sail away when the wind changes, their ship is destroyed by a thunderbolt and they all drown. At this point the story has much in common with a story told about the infant Dionysus, child of Zeus and posthumously of Semele, who used to play with the Titans.[1] The Titans decided to cut him up and eat him, so they put the pot on the stove and got to work. The meat was cooking and the Titans had tasted the flesh of the god when Zeus smelt the savor rising to heaven and pulverized the Titans with a thunderbolt. The child was miraculously put back together again and from the dust of the Titans the human

---

1. *Thomas Taylor the Platonist*, op. cit., Olympiodorus, pp 409–412.

race was made. Because we are made from the dust of the Titans who had tasted the flesh of Dionysus, there is in us an element of the god. In the same way the companions tasted the immortal cattle. I have suggested already that the corpses on the island of the Sirens and the victims of Scylla correspond to the souls in Plato's *Myth of Er* who voluntarily re-enter the planetary realm on their way to another mortal life and death. As much is true of this last remnant of Odysseus' company who are destroyed for eating the Cattle of the Sun. For this crime they are precipitated into a mortal life, a life which is really the immortal life of the sun god's cattle. Every creature on earth has its natal day; they are all the divine energy of their birthdays, released in partial, mortal forms on earth.

This is a difficult notion to grasp. We are to imagine the deaths of Odysseus' companions as being births into mortal life, and that the days of the year are more real and actual than we are ourselves. Our characters as creatures on earth are really their characters as members of the annual cycle. To escape the power of the planets, to refuse to return to a mortal life on their terms, it is as necessary to leave unharmed the Cattle of the Sun as it is to escape the Sirens, Scylla and Charybdis. To kill and eat of the cattle or sheep binds one irrevocably to a day or night of the year, so that ones own power is overcome and wholly submerged by the immortal power of the god in mutilated form. Odysseus does not taste the meat and so he alone is able to swim away to safety. But still we ask: how can we ourselves be considered less real than the calendar periods, such that our lives are no more than the remnants of theirs? In the context of the interpretation we have given of the Sirens, Scylla and Charybdis, this reading of the Cattle of the Sun provides more detail of our submission to the heavenly powers, in this case in consequence of an act of sacrilege against them. The frightful power of Scylla and Charybdis is modified here and the adventure is a seduction in the manner of the Sirens rather than a violent assault. But the weird results of killing the immortal cattle, the crawling hides and lowing meat, show a new grimness in our relations to the solar system, of a kind which the other adventures only suggest.

The interpretation of the penultimate journey which I have given is far from exhaustive and is often self-contradictory. But it does provide an avenue of approach to the four adventures which links them as closely in meaning as they are linked in narrative, and which converts them from a series of frightening fairy tales into an analysis of the solar system. Homer's scientific interests are tempered by his love of story and dramatic effects, so that the scientific facts are released in imaginative possibilities which move our feelings. And, as with the location of Hades at the South Pole, we find here, intermingling with the scientific account of the planets, a psychic and spiritual dimension which seems to us of another order altogether. But for all that, the system of the planets still looms behind Homer's stories, and I have suggested that Plato, with his broad hints and allusions to Homer, is telling us this in the *Myth of Er*. But if this is Homer's subject in these adventures, then his poem takes on a rather different appearance. It becomes knowledge, however partial, of the same kind that we think knowledge, just as the accounts of Laestrygonia and the rest adumbrate what we regard as scientific geography. This manoeuvre transforms a poem which is loved as one of the most fantastical we have, into a useful primer to some of the hard sciences. And in many cases it is just what once appeared most fantastical which has become most scientific.

In claiming that these adventures of Odysseus take him past the planets, I do not wish to imply that Homer's *Odyssey* is a space Odyssey after Arthur C. Clarke. There is no suggestion, I believe, that travelling beyond the solar system in a spaceship would confer physical immortality on the astronauts. We may take it rather that to escape the limitations of mortality was to escape the planets. Mortal creatures are, on this account, spiritual, psychic and physical complexes, the structure of which is organized after the largest structure of all, the universe itself. To transcend the planets is to leave this complex behind. This is what Odysseus' reaching of Ogygia means, however sensual the ageless immortality which Calypso offers him. For it is a principle of such transcendence that all the forms of the condition transcended are reinstated at the higher and more perfect level from which the condition transcended derived them. This is what Odysseus achieves. When we add the adventures of the penultimate journey to the ones discussed in the first chapter we begin to see just how well travelled Odysseus was. Not only did he cover all the latitudes of the earth's surface, but all the regions of the planets too. His journey is not a journey but the journey, since it covered all the ground possible. By virtue of his journey Odysseus is a universal man. By the time he reaches Calypso he is scarcely a man at all, but an immortal.

Apart from the paradox that Odysseus enjoys a sensuous life with Calypso, when he is on this account beyond a mortal existence, there is also the paradox that in dying on the journey, the companions correspond to the souls about to be reborn in the *Myth of Er*. Odysseus survives to live with Calypso while his companions, eaten or drowned, are dead. But the distinction here is not between being alive and being dead, but between being immortal and being mortal. The companions who die are clearly mortal while Odysseus is offered immortality by Calypso. The nature of this immortality is hard to conceive but it is at least ageless. We may compare it to what Proteus foretells to Menelaus, how Menelaus will be taken to the Elysian Plain by immortals, where life is easiest for men.[1] By surviving all those adventures Odysseus has gained immortality if he should choose to accept it. That he chooses not to accept it does not alter the spiritual achievement. Just as the delights of life with Calypso reinstate the pleasures of a mortal existence at a higher and permanent level, so the achieving of immortality confers upon the achiever the power to return to a mortal condition at will. For the higher state necessarily controls the lower states which derive from it. By transcending mortality Odysseus is empowered to return to mortality if he should wish to do so. Even so, Odysseus grieves for many years on Ogygia until he is released.

But if Homer intended the adventures of the penultimate journey to represent a passing beyond the planets, why did he not make his meaning clearer? To this question there are several possible answers. It may be that the *Odyssey* was deliberately contrived to bear two different levels of meaning at once, an esoteric and exoteric reading, and that the planetary interpretation was part of the esoteric reading. In order to preserve the surface of the exoteric reading, the planetary significance of the adventures is left in the background. That

1. *Odyssey*, IV, 563–569.

games and amusements have occult meanings, largely or completely unimagined by their players, is not uncommon, and this may be the case with the *Odyssey*. The genius of Homer as a teacher might have consisted precisely in this: that he could present his lessons in such a way that even those who had not understood their deeper significance were enthralled. His lessons were passed on for millennia, to the secret delight of those who could see beyond their surface. It is certainly true that the *Odyssey* at the simplest level is so fine a story that it has survived, however little understood. I have suggested that Plato's *Myth of Er* is, in part, an interpretation of this part of the *Odyssey* and that Plato makes this clear enough by his direct references to Alcinous and the Sirens. If this is true, then our difficulties in seeing the planetary meaning here may spring from our own cultural circumstances rather than from Homer's obscurity. We are not used to thinking that poets cover the same ground as scientists, since we have dissociated the processes and discoveries of rational scientific inquiry from the arts and letters. Plato did not do this. This development makes it very difficult for us to treat poems with the seriousness which our ancestors accorded them, since the medium of poetry has long since ceased to be a vehicle for scientific understanding. And with this shift comes a certain condescension towards the ancients, and a readiness on our part to accept that since those people lacked our science, they were perfectly satisfied with stories of six headed monsters and immortal cattle. We easily believe that they required nothing more to satisfy their desire to know than fairy tales, and that this great poem was an adventure story.

So the reading of the penultimate journey which I have given is not necessarily esoteric, even if we grant that Homer meant his story to be read this way. For it may be that this reading was clear enough to the Greeks in general, though it is difficult for us because we can no longer combine science and poetry in the manner required. In Plato's *Myth of Er* we hear, it seems to me, a clear enough account of the same cosmos as I have discovered in Homer's stories, and hints and allusions enough to force us to the comparison. But it is also true that there is very little in the earliest writings of the Greeks which is clearly and unambiguously concerned with the planetary powers. Is this because they were not concerned with such things?

Or because, though they did contemplate them, they concealed their thoughts about them? Or have vital works on these matters not come down to us? Or might it even be that the movements of the planets were so universally realized in the forms and practices of their culture, that we cannot see them because they are everywhere? The Greek astronomer Aratus confessed that he lacked the daring to give an account of the planets after describing the fixed stars.[1] The Roman astronomer Manilius seems to have followed Aratus in this and omitted any account of the planets. If the twelfth book of the *Odyssey* is an account of the planets, then it appears to have been left, with Plato's myths, to stand on its own.

---

1. *Hymns and Epigrams,* tr. A.W. Mair (London: Loeb Classical Library, 1921), Aratus, 'Phaenomena', 454–461.

3

CHRISTMAS IN ITHACA

THE LAST LEG of Odysseus' journey home is from Calypso's island to Ithaca via Phaeacia. The adventures on this last journey are as closely linked with each other as those we have just considered in the last chapter. Since, however, Homer has chosen to narrate Odysseus' arrival and most of his experiences in Phaeacia at the beginning of his account of Odysseus, but does not tell of his arrival in Ithaca until much later, it is easy to overlook the similarities which bind together this last phase of the journey. On the other hand, the separation of the arrival in Phaeacia from the arrival in Ithaca in Homer's narration, at the beginning and end of his account of Odysseus' journey, allows a certain symmetry to his account of the whole journey which begins and ends in so similar a way. It takes Odysseus nineteen days to reach Phaeacia from Ogygia, as it takes him nineteen years to return to Ithaca. Odysseus arrives in Phaeacia in the aftermath of a terrible storm which has wrecked his raft, while he arrives in Ithaca in a deep sleep on the magical boat of the Phaeacians. But Odysseus is saved from the storm by the magical scarf of Leucothoe, and the Phaeacian boat is as surely lost on its return to Phaeacia as the raft on which Odysseus sailed most of his way there. In each case Odysseus wakes up alone to find himself on an unknown shore where he is helped by a beautiful virgin accompanied by her attendants: in Phaeacia it is Nausicaa and her maids, in Ithaca Athene and the Nymphs called Naiads who inhabit the cave at which he is landed. On waking in Phaeacia Odysseus seriously wonders whether the cries of Nausicaa's maids which wake him are the cries of Nymphs, and it is Athene who sends Nausicaa to meet him there.

In both places Odysseus is advised by his beautiful acquaintance on how to cope with the new situation in which he finds himself, and the two situations are very similar. There is a town some distance away, and a palace, and the country is potentially hostile. It is as though Odysseus' landing in Phaeacia is a dry run for his landing in Ithaca. The unusual number of parallels between these two landings aroused the interest of William Blake in the last years of his life and led him to compose a picture, "The Sea of Space and Time", in which he depicts the arrival of Odysseus in Phaeacia and his throwing back of the veil to Leucothoe against a backdrop of the Cave of the Nymphs in Ithaca.[1] In this way Blake conflated the two landings in a single scheme. Standing beside Odysseus as he throws back the scarf is the figure of a beautiful woman clad in a diaphanous robe and pointing upwards to heaven. This figure is both Nausicaa and Athene who appear indistinguishably as the guardian goddess or daemon, appointed to assist Odysseus in each of these new births on the shores of life. As the climactic moment of Odysseus' arrival in Ithaca draws close, it seems as though everything conforms and conduces to it. Distinct events in widely separate places seem, as it were, to come into phase with each other, so that the arrival in Phaeacia is a premonition of the arrival in Ithaca.

The arrival in Ithaca is one of the two climaxes of the *Odyssey*. The other is the battle with the suitors. Odysseus arrives in Ithaca in a sleep like death on board a ship which can go anywhere and return in a single day, and which charges through the sea like a four-horsed chariot. It is an image of stillness in movement. And just as Homer describes the speed of the ship he pronounces an epitaph over the sleeping Odysseus.

> Thus she sped lightly on, cutting her way through the waves and carrying a man wise as the gods are wise, who in long years of war on land and wandering across the cruel seas had suffered many agonies of spirit but now was lapped in peaceful sleep, forgetting all he had once endured.[2]

This moment and Odysseus' first moments in Ithaca are the pivot on which the return of the *Odyssey* swings. As that ship cuts the

---

1. 'The Sea of Time and Space' is also known as the Arlington Court Tempera.
2. *Odyssey*, XIII, 88–92.

waves there is a sense of closure as the saga of Odysseus' adventures away from Ithaca comes to its end. But this is only temporary, a lull in the eye of the hurricane, before he is caught up again into his struggle with the suitors. That his fortunes are on the turn is made clear by his encounter with Athene face to face. Athene's manner towards Odysseus on this occasion, her divine condescension towards him and the intimacy of their conversation, makes the moment one of the most moving theophanies in ancient Greek literature. Odysseus is home at last, not only in Ithaca, but in the presence of the goddess. He had addressed Nausicaa as a goddess on the shore of Phaeacia. Here, when at length he recognizes Athene, he upbraids her for her long absence from his side.

The centrality of Odysseus' return to Ithaca is signified also by its place in the poem, at the beginning of the thirteenth book, in the middle of the twenty-four. It is emphasized further by the time at which it occurs and its location, the Cave of the Nymphs. Professor Gilbert Murray has suggested that Odysseus' return to Ithaca in the twentieth year of his absence occurred at the time of the winter solstice and the dark of the moon, at the very end and very beginning of a Metonic cycle of nineteen years, dated from the winter solstice.[1] This cycle of nineteen years is the length of time taken by the moon to return to the same position in relation to the sun, thereby harmonizing the lengths of time taken by the sun and moon in their yearly and monthly courses. According to Professor Murray, this conjunction of sun and moon after nineteen years corresponded exactly to the reunion of Odysseus and Penelope after the same period. In this suggestion Professor Murray has been followed by Joseph Campbell.[2] This view of the time when Odysseus returned rests on the assumption that astronomical matters were of concern to Homer, and this is attractive. On the other hand, Professor Murray's suggestion goes well beyond the evidence in supposing that Homer knew of the Metonic cycle.

1. Gilbert Murray, *The Rise of the Greek Epic* (Oxford: Clarendon Press, 1924), p211.

2. Joseph Campbell, *Occidental Mythology* (Middlesex: Penguin Books, Ltd., 1976), pp163–164.

It is clear from the *Odyssey* that Odysseus returns to Ithaca in the winter. It is a time of large fires in the hearth, heavy cloaks and morning frosts, a time when many blankets are needed on the bed. Odysseus, incognito, explains to Eumaeus the swineherd and to Penelope:

> This very year Odysseus will be here. Between the waning of the old moon and the waxing of the new....[1]

The word translated here as 'year' has been taken to mean 'month' or even 'day', another indication, perhaps, of how just these passages aroused the greatest interpretative fervor. We are also told that the battle in the hall takes place on the feast day of the archer god Apollo.[2] There is an aptness in Odysseus' returning between the two moons which would be increased if his return coincided with the changing of the year also, and it is true both of ourselves and of the Romans after Julius Caesar that our new year begins about the time of the winter solstice. But this was by no means the universal practice in ancient Greece. The Athenian year began from the summer solstice, not the winter, and the Spartan year began at the autumn equinox. On the other hand, the feast day of Apollo may well signify the New Year with its ceremony of shooting the arrow through the twelve axeheads. The successful completion of this task would represent the sun's passage of the twelve zodiacal constellations and mark the end of the year. The mention of the moon's phase is also relevant to the dating of the year because the Greeks tended to date the year from the time of the last appearance of the old moon after the solstice or equinox.

We may then tentatively propose that Odysseus returns to Ithaca at the time of the winter solstice and the dark of the moon, from which latter time the Ithacan New Year was reckoned. But even this proposal, tentative as it is, falls short of Professor Murray's claims about the Metonic cycle. We do not know exactly when Odysseus left Ithaca to go to Aulis and Troy and we cannot therefore measure the exact length of his absence. It is possible that Odysseus left Ithaca at

---

1. *Odyssey,* xiv, 161–162; xix, 306–307.
2. Ibid., xx, 278; xxi, 258–259.

the time of the winter solstice, but we are not told that he did and it is improbable that he would have departed on a military expedition at this time of year. But the strongest objection to Professor Murray's thesis is that it contradicts what we know about the evolution of the Greek calendar. The synchronizing of the solar and lunar cycles is difficult and the Greeks appear to have tried two, four and eight year periods, until a hundred years after Meton's publication of the nineteen year period in 432 BC. In the fourth century the Athenians adopted this system, dating the period from the last of the moon after the summer solstice. But Homer lived during or before the eighth century BC and his poems were well known from the sixth century onwards. It is unlikely that the Greeks would have delayed the introduction of the nineteen-year cycle as the basis of their calendar if it was known from Homer and had served as the astronomical basis of the *Odyssey*. For in that case they would have had the strongest religious as well as practical reasons for adopting it.

That Odysseus returns between the old and new moons is suggested in other ways in the poem. The first night which Odysseus spends in the hut of Eumaeus is specifically described as moonless, though it is also a dirty night.[1] Homer describes the very moment when Odysseus arrives at Ithaca on the Phaeacian ship in the following terms:

> When the brightest of all stars came up, the star which often ushers in the tender light of dawn, the ship's voyage was done and she drew near to Ithaca.[2]

At first sight it appears that the 'brightest of stars' is Venus, the morning star, which often rises just before the sun. In that case we might attempt to date Odysseus' return by consulting the ephemeris for years in which Venus rose as the Morning Star at the time of the winter solstice. But at the end of each month the moon, too, rises just before the sun and the Athenians, as we have seen, dated the first day of the first month and the first day of the new year from the last appearance of the old moon after the summer solstice. Gilbert

---

1. *Odyssey*, XIV, 457–458.
2. Ibid., XIII, 93–95.

Murray and Joseph Campbell take it that the feast day of Apollo marked the winter solstice and that in the year of Odysseus' return this day happened to fall on the day of the dark of the moon. Since, so they claim, this occurs only once every nineteen years and since Odysseus and Penelope were separated for nineteen years, therefore the day of Odysseus' return marks the beginning of a Metonic cycle. But the coincidence of the feast day of Apollo with the dark of the moon does not entail a coincidence of the winter solstice with the dark of the moon. For the Ithacans may have adopted the same practice as the Athenians, of transferring the celebration of the solstice to the last day of the old moon which followed it. In that case this feast would fall on the day of the dark of the moon every year. Of course, on this interpretation, we must forego the elegant explanation for Odysseus' nineteen-year absence. What is needed to save this is an explanation of why this cycle was not adopted as the basis of the calendar long before the fourth century if it was known to Homer.

There are two other passages towards the end of the *Odyssey* which may bear upon the time of the year when Odysseus returned. The first of these is the vision of Theoclymenus shortly before the battle of the hall. At this point Athene turns the wits of the suitors who laugh with alien lips and their meat is dabbled with blood. Theoclymenus perceives this:

> 'Unhappy men,' he cried, 'what blight is this that has descended on you? Your heads, your faces, and your knees are veiled in night. There is a sound of mourning in the air; I see cheeks wet with tears. And look, the panels and the walls are splashed with blood. The porch is filled with ghosts. So is the court—ghosts hurrying down to darkness and to Hell. The sun is blotted out from heaven and a malignant mist has crept upon the world.' They laughed at him. . . . [1]

This sense of the darkness gathering around the suitors accords with the season since it is the darkest time of year. Of course, the darkness in the hall is visionary and is seen by the prophetic eye in broad daylight, but it also realized the winter solstice as a moral and

1. *Odyssey,* xx, 351–357.

human phenomenon, as though the darkness of the season and the presence of the ghosts were caused by the atrocity of the suitor's actions and the nearness of their doom. We note again the connection between the appearance of the dead and the south, in this case the southernmost position of the sun's course. A similar effect is gained when Athene delays the sun's rising on the night of Odysseus' and Penelope's reunion. Athene appears to prolong the night that they may pass beyond their grieving at the lost years of their marriage. Here the length of the night, the darkness of the season, is realized as a human phenomenon in a positive and joyful way, and we are reminded of all those love poems in which passionate lovers abjure the sun not to rise so that they may continue to enjoy the beloved. In these ways the phenomenon of the shortest days of the year becomes a direct expression of human deeds and human feelings. The events in Ithaca acquire a significance so tremendous that they arrange the very courses of the heavens.

Odysseus returns to Ithaca in winter, at the darkest time of the year, and we have assumed that he tells Eumaeus and Penelope that Odysseus will return this very year, between the waning and the waxing of the moon. It may be the last of the moon when Odysseus first arrives in Ithaca and the dark of the moon when he spends his first night with Eumaeus. His advice to Eumaeus and Penelope stresses the imminence of Odysseus' arrival, which has in fact already happened. So it is likely that the year will end with the month which is already almost over. The successful passage of the arrow through all twelve axeheads also suggests the completion of the year. But if the Ithacan year, like the Roman year and our own year ends and begins at about the time of the winter solstice, yet the Ithacan year seems in some ways to be related to lunations and not just solar events, unlike the Roman year and our own which are exclusively solar. That there is a feast day of Apollo at this time also recalls our own and the Roman calendar which sets the last days of the year aside as a period of holidays, during which the birthday of a god or the birthdays of gods are celebrated. It appears that just as doorways and thresholds were considered sacred because they were neither inside nor outside but somehow beyond space in a special space of their own, so this period between the old and the new years was betwixt and between, neither one thing nor the other and therefore beyond time in a special time of its own. The point is made by the name of our own and the Roman first month, January, the month of the doors.

In the case of the Roman calendar, the beginning of the New Year at the winter solstice was immediately preceded by a festival called the Saturnalia, which began as a one day festival of the god Saturn but was gradually increased over the centuries until it occupied the full week before the solstice. It is generally believed that this Roman festival of Saturn derived from a Greek festival, the Croniadae, in honour of the same god. The festival of the Saturnalia is of particular relevance to our understanding of the time and nature of Odysseus' return to Ithaca. For it is described at length by the philosopher Porphyry in his essay on Homer's Cave of the Nymphs. Porphyry lived in the third century CE and was a disciple of the Platonic philosopher Plotinus. His substantial essay on the Cave of the Nymphs is the fullest example which has come down to us of the allegorical method of interpreting Homer among the ancients. Porphyry discusses Odysseus' return in the following terms:

> For the Romans celebrate their Saturnalia when the sun is in Capricorn, and during this festivity the servants wear the shoes of those who are free, and all things are distributed among them in common; the legislator intimating by this ceremony, that those who are servants at present, by the condition of their birth, will be hereafter liberated by the Saturnalian feast, and by the house attributed to Saturn, i.e. Capricorn; when reviving in that sign, and being divested of the material garments of regeneration, they return to their pristine felicity, and to the fountain of life.[1]

According to Macrobius in the fourth century CE in his work called the *Saturnalia*, the festival included a feast at which the masters served the slaves.

Porphyry, we may assume, was impressed by the coincidence of the time of year when Odysseus returned with the date of the Saturnalia. He supposes, I think, that Odysseus and the suitors experience something of the same reversal of roles which characterized

---

1. *Thomas Taylor the Platonist*, op. cit., Porphyry, p312.

the Saturnalia. The king goes in disguise as a beggar and the slaves Eumaeus and Philoetius act the part of warrior heroes. According to Porphyry the festival symbolized the time when slaves and free people would share in the great equality of death, since the winter solstice marks the death of the year. It is an uplifting thought that the festival restored the essential equality of humankind for a few brief hours in that time which is no time at the very end of the year. But it is difficult to assess how far Porphyry was right to suppose that the festival bore comparison with the manner of Odysseus' return. The Roman Saturnalia was, like our Christmas, a time of universal goodwill and festive mirth. There is a kind of mirth in the suitor's doings in the *Odyssey* but it is tinged as Theoclymenus saw, with a desperate sadness. Ctesippus throws an oxhoof at Odysseus' head, who moves so that it misses and hits the wall. At that moment Odysseus smiles to himself, a grim and bitter smile. This is not the good humour of the Saturnalia but an inversion of it. And there is an even more important difference between the Saturnalia and the return of Odysseus. During the festival of the Saturnalia neither slave nor criminal could undergo punishment. But when Odysseus returns, a hundred suitors are killed. As Odysseus describes it, in the same grim and bitter way, after stringing the bow and shooting the arrow:

> But the time has come now to get their supper ready, while it is light, and after that to pass on to the further pleasures of music and dancing, without which no banquet is complete.[1]

After the slaughter Odysseus orders that the sounds of merrymaking be maintained to conceal from outsiders what has happened in the palace.

In the case of the city as a whole, Odysseus' return is not merely a resumption but a restoration of the kingship. The wheel has gone round half a turn since his departure from Troy. He had left Ithaca after some years of governing it well. By the time of his return, his city has degenerated. Now it is in the hands of the suitors whose outrages have grown ever worse and more frequent. If Phaeacia is

---

1. *Odyssey*, XXI, 428–430.

the model of good government in the *Odyssey,* Ithaca is the very opposite. The influence of good king Odysseus has waned to nothing. The suitors have conspired to murder Telemachus, to obliterate the line of Laertes and Odysseus. Teiresias has foretold to Odysseus that if he manages to return he will take vengeance on the suitors and live to a ripe old age among a prosperous people. So the point of his return marks the end of the decline and the beginning of a new prosperity for Ithaca. From this point of view we may compare the conduct and mentality of the suitors with Plato's account of the tyrant in the later books of the *Republic.*[1] For Plato, too, any human society was a natural organism which had its own cycle of growth and decay and he describes the disorder of the last stages in this process in great detail. We may regard the Saturnalia too as an inverted hierarchy, as a miniature representation at the end of each year of the chaos which overtakes a city in the last phases of political dissolution at the end of a much larger cycle. The New Year becomes a restoration of the government, the vesting of a new energy in the political order to maintain the state for another year. There is a parallel to the renewal of the ritual kingship once a year by mortal combat between the incumbent and any challenger.

In the human life cycle the last days of the old year correspond to extreme old age and the beginning of the new year corresponds to birth. Of the first type Saturn, Old Father Time and Father Christmas may serve as examples; of the second type the baby Jesus, the Christ child. In Christian symbolism the encounter between the old and the new is beautifully represented in the adoration of the Magi. In the *Odyssey* Laertes, Odysseus' father, is the figure of the declining year who in grief for his wife and his son has withdrawn further and further from the human world, working and sleeping in the fields, taking no care of his own person. To these griefs must be added the violence of the suitors and Telemachus' journey to seek news of his father, since the old man fears now for his grandson's safety as well. He is in the deepest despair and when Odysseus convinces him that he is back, the old man literally expires, though the

1. Plato, *The Republic,* tr. Desmond Lee (Middlesex: Penguin Books, Ltd., 1974), pp 562 ff.

word which Homer uses means here to faint or swoon. And then there is a rejuvenation, aided by Athene, which culminates in Laertes' killing of Eupeithes, Antinous' father, in the last lines of the poem. In this way Eupeithes becomes the sacrificial representation of the old year, a substitute for Laertes who has renewed himself. Laertes, as it were, steps over the gulf or chasm between the old and the new.

What, then, of the newborn baby who represents the New Year? To find this birth we must move on from the time of Odysseus' return to its exact location, the Cave of the Nymphs:

> Now in that island is a cove named after Phorcys, the Old Man of the Sea, with two bold headlands squatting at its mouth so as to protect it from the heavy swell raised by rough weather in the open and allow large ships to ride inside without so much as tying up, once within mooring distance of the shore. At the head of the cove grows a long-leaved olive-tree and near by is a cavern that offers welcome shade and is sacred to the Nymphs whom we call Naiads. This cave contains a number of stone basins and two-handled jars, which are used by bees as their hives; also great looms of stone where the Nymphs weave marvellous fabrics of sea-purple; and there are springs whose water never fails. It has two mouths. The one that looks north is the way down for men. The other, facing south, is more divine; and as immortals come in by this way it is not used by men at all. It was here that the Phaeacians put in, knowing the spot. [1]

Here is Porphyry's account of the stone looms on which the Naiads weave their purple garments:

> And what symbol is more proper to souls descending into generation, and the tenacious vestment of body, than as the poet says, 'Nymphs weaving on stony beams purple garments wonderful to behold'. For the flesh is generated in and about the bones, which in the bodies of animals may be compared to stones. On which account these textorial instruments are fabricated from stones alone. But the purple garments plainly appear to be the flesh with which we are invested; and which is woven as it were and grows by the connecting and vivifying power of the blood, diffused through every part. Besides, purple garments are tinged with the blood of animals; and

---

1. *Odyssey,* XIII, 96–112.

flesh is produced and subsists from blood. Add too that the body is a garment with which the soul is invested; a circumstance indeed wonderful to the sight, whether we regard its composition, or consider the connecting band by which it is knit to the soul.[1]

Porphyry supposes that the stone looms and purple garments are bones and flesh. Inside and outside the womb, the flesh is woven on the bones, since even in old age a cut finger heals itself. While we are alive, on this account, the Nymphs are always busy at their work of repairing our bodies. But this is never more true than in the womb where the body is shaped for the first time. Homer describes these Nymphs as in a cave, a cave through which perpetual waters flow like the waters of the womb. But if the cave is the womb, then the two headlands are the outspread legs and thighs. These headlands are sheer cliffs towards the sea but sloping down on the inside, according to some translators. The double doors of the cave's entrances, may be the vulvae. The leaves of the long-leaved olive tree are the pubic hairs, for the leaves of plants, the hairs of animals, the feathers of birds and the scales of fish are the same thing, according to Empedocles.[2]

The Nymphs are Naiads, water Nymphs, semi-divine beings who live in the cave on the cove which is sacred to the old man of the sea. If these Nymphs weave the flesh on our bones like garments, then there is another correspondence between them and the maidens who attend Nausicaa. Nausicaa's maidens place a cloak and a tunic beside Odysseus as he washes the salt from his body. The Nymphs of the cave are very Ithacan Nymphs, to whom Odysseus has sacrificed many hecatombs in times past. Now he vows more to them if Athene should save him and Telemachus.

Why should a sacrifice to the Nymphs be a way of giving thanks to Athene? The relationship between Athene and the Nymphs of Homer's cave corresponds to certain features of Athene's cult in Athens. There, at the Panathenaea, Athene was ritually presented with a robe woven by maidens. This robe, moreover, was begun each year at the festival of the Chalceia which occurred nine months

1. *Thomas Taylor the Platonist*, op. cit., Porphyry, p305.

2. *Ancilla to the Presocratic Philosophers*, ed. Kathleen Freeman (Oxford: Basil Blackwell, Empedocles, 1971), frag. 82.

before the Panathenaea.[1] Our knowledge of the Panathenaea comes
from a later period than the *Odyssey,* but Homer describes in the
Iliad a ceremony very similar to the Panathenaea when the Trojan
women present a robe to Athene in her temple with the help of her
priestess Theano.[2] From this point of view we may regard Athene as
the deity who presides over the weaving of the Nymphs, as she is the
deity in whose honour the robe was made yearly in Athens, and to
whom it was presented in the *Iliad.* This correspondence between
Homer's Athene and Athenian cult may be taken one step further.
For it is Athene who suggests to Nausicaa that she should wash the
clothes of her family, and in Athens at the festival of Plynteria in
spring, the robe of Athene was ritually washed. At this level, too,
there is an equivalence between Nausicaa's maidens and the
Nymphs of the cave. Their activities, weaving and washing, are the
complementary rituals of Athene's Athenian worshippers. If Herodotus was right to say that Homer and Hesiod instituted the religious forms of Greece,[3] then the Athenians may in part have derived
these forms of their worship from these passages in the *Odyssey.*

Homer does not begin his account of the cave's contents with his
description of the weaving Nymphs. The first part of his account
describes the stone basins and jars where bees leave a store of honey.
These bees correspond to the Nymphs who weave on looms of
stone, but I see no need to read the bees symbolically. Porphyry
writes at length about the significance of honey, suggesting among
much else that it stands for the sweetness of sexual union which
draws souls down into life.[4] The honey may also be compared to
human semen; the basins and jars are womblike containers. But this,
I think, is to miss the force of Homer's point. The cave, he has told
us, is sacred to the Nymphs called Naiads. In the cave there are
basins and jars, which we would expect as votive offerings to the
Nymphs to whom the cave was known to be sacred. These basins

---

1. K. Kerenyi, *Athene: Virgin and Mother in Greek Religion* (Dallas: Spring Publications, Inc., 1988), p 29.
    2. *Iliad,* vi, 269–311.
    3. *Herodotus,* op. cit., ii, 53.
    4. *Thomas Taylor the Platonist,* op. cit., Porphyry, pp 306–308.

and jars are made of stone like the looms. In these basins and jars the bees store honey. Honey is sometimes a sacrificial offering made by humans to gods and here in the vessels offered to the gods, the bees of their own accord leave honey, as though they, too, wished to make their offering to the Nymphs. Charming though this is, we should not forget that Homer has said that the bees store their honey there, not sacrifice it. They will be back when the time demands.

We enter the cave, pass the votive offerings of humans filled with the offerings of bees, and come to the Nymphs at their weaving. Are they more like the people who carved and sacrificed the basins and jars or like the bees who gave the honey? Insofar as they are anthropomorphic and engaged in the human task of weaving, they are obviously human. But insofar as their work is the creation of our bodies at a level of existence beyond our conscious volition or understanding, in a manner which is marvellous in its beauty and organization, the Nymphs are more like the bees. The bees are like the Nymphs at their work of creation and preservation about our limbs. The stone vessels and the stone looms are images of the perdurable material on which they do this work aeon after aeon.

Why does Odysseus return to this point in Ithaca? This place symbolizes the womb and Homer represents the return as a birth or rebirth. This answers our requirement that there be a new born human being to correspond to the new year as Laertes and Eupeithes correspond to the old year. This new birth is not the birth of a baby but a fully mature man, and he is not born of a woman but of the land itself, from the cave which is also a womb. We might claim that Odysseus, like the ancient Athenians, is earth born from the rocks and stones of his native land. It is notable that Odysseus does not at first recognize Ithaca, since Athene has cast a mist about him so that he will not be seen or recognized by any of his country folk. This mist hides Ithaca from Odysseus just as it hides Odysseus from Ithaca. Then Athene appears and talks to Odysseus. First she is like a shepherd prince and then a tall and beautiful woman. As shepherd she tells him that he is in Ithaca but it is not until after her transformation into a woman and a long conversation that she moves the mist aside, saying:

And now, to convince you, let me show you the Ithacan scene. Here is the harbour of Phorcys, the Old Man of the Sea; and there at the head of the haven is the long-leaved olive-tree with the cave near by, the pleasant shady spot that is sacred to the Nymphs whom men call Naiads. Over there you can see its vaulted roof—it will put you in mind of many a solemn sacrifice you have made there to the Nymphs—while the forest-clad slopes behind are those of Mount Neriton.[1]

In his essay on the cave Porphyry quotes an otherwise unknown half line from the fifth century poet and philosopher Empedocles. According to Porphyry, Empedocles said that the psychopompic powers say:

We have come under this roofed cave.[2]

Porphyry quotes this half line of Empedocles to illustrate how the ancients conceived of caves, and he goes on to mention Plato's simile of the cave in the *Republic*. The point of his remark about the psychopompic powers seems to be that as the soul is born into the world, or perhaps the womb, its guiding powers announce to it that it has now come into a confined space. These psychopompic powers remind us of the guardian genii in the *Myth of Er* who are appointed to souls after they choose their lots, to guide them through life and ensure that they fulfil their self selected destinies. The words quoted from Empedocles fit a context very like that of these words of Athene to Odysseus, which suggests that Athene is the psychopompic power of Odysseus, his guardian genius. Athene leads Odysseus into the cave and shows him where to leave his treasures. So the encounter between Athene and Odysseus on the shore of Ithaca may be an encounter between a man and his guardian genius, like the appointing of a guardian genius to each soul just before its birth in the *Myth of Er*. In the *Myth of Er*, however, the souls drink of the waters of Unmindfulness after that moment but before they are born. In the *Odyssey* Odysseus meets his genius face to face and in full knowledge at his second birth. In the line of Empedocles, the psychopompic

1. *Odyssey,* XIII, 344–351.
2. *Thomas Taylor the Platonist,* op. cit., Porphyry, p302.

powers appear to speak directly to the soul at the moment of birth or conception, in order to orient the soul in its new environment. This is what Athene does to Odysseus.

The meeting between Odysseus and Athene by the cave of the Nymphs is the most extended theophany of a god to a mortal in Homer's work. The contrast between this meeting and Odysseus' meeting with Nausicaa is engaging. To Nausicaa Odysseus conducts himself as to a goddess and their relationship up to the moment of their parting is chaste and reverent on both sides, though each feels the attraction of the other. With Athene, on the other hand, the conversation is more relaxed, though Athene is the goddess, not Nausicaa. Athene says she could never desert Odysseus because he is the best of men in deliberation and lying tales as she is the best of gods in trickery and cunning thoughts. Here is the essential connection between them. And she says this to him as she strokes him with her hand, the hand of a tall and beautiful woman. With Nausicaa, Odysseus forbore from embracing her knees and refused to be bathed by her maids. With Athene, Odysseus is entirely at ease.

In his essay on the life of Plotinus, Porphyry tells a story about Plotinus' presiding spirit:

> In fact Plotinus possessed by birth something more than is accorded to other men. An Egyptian priest who had arrived in Rome and, through some friend, had been presented to the philosopher, became desirous of displaying his powers to him, and he offered to evoke a visible manifestation of Plotinus' presiding spirit. Plotinus readily consented and the evocation was made in the Temple of Isis, the only place, they say, which the Egyptian could find pure in Rome.

> At the summons a Divinity appeared, not a being of the spirit-ranks, and the Egyptian exclaimed: 'You are singularly graced; the guiding-spirit within you is not of the lower degree but a God.' It was not possible, however, to interrogate or even to contemplate this God any further, for the priest's assistant, who had been holding the birds to prevent them flying away, strangled them, whether through jealousy or in terror. Thus Plotinus had for indwelling spirit a Being of the more divine degree, and he kept his own divine spirit unceasingly intent upon that inner presence. It was this preoccupation that led

him to write his treatise upon *Our Tutelary Spirit,* an essay in the explanation of the differences among spirit-guides.[1]

What was revealed of Plotinus in the magical ceremony is revealed of Odysseus in his encounter with Athene. Odysseus, too, is singularly graced in having so powerful a god for his presiding spirit. This is the second indication which we have that Odysseus is very unusual, if not unique, among mankind. The first indication was his being offered immortality by Calypso.

Now for the last of what Homer say of the cave of the Nymphs.

> It has two mouths. The one that looks north is the way down for men. The other, facing south, is more divine; and as immortals come in by this way it is not used by men at all. It was here that the Phaeacians put in, knowing the spot.[2]

The Phaeacian ship arrives at the southern entrance. Odysseus later follows Athene into and out of the cave when she leads him in and shows him where to leave his treasure. Homer does not say that the northern entrance is inaccessible to gods, so the two of them may have entered through the northern entrance. But we are not told that they do so, where we are told that the southern entrance is the path of immortals, to whom Athene clearly belongs. But if Odysseus uses the southern entrance to enter the cave, then he is an immortal too, for human beings do not use that entrance. But we know that Odysseus is not an immortal because we know that he will die. Teiresias prophesies his death in detail when Odysseus raises the dead.

There are two exceptions to this entrance rule. These exceptions are set out by R. Guénon in his work on the symbolic zodiac of the Pythagoreans:

> The 'gate of the gods' cannot be an entry except in the case of voluntary descent into the manifested world, either by a being already 'delivered', or by a being representing the direct expression of a supra-cosmic principle. But it is obvious that these exceptional cases are not part of the 'normal' processes that we are considering here.[3]

1. Plotinus, *Enneads,* tr. S. McKenna (London: Faber and Faber, 1969), Porphyry, *Life of Plotinus,* 10, p8.
2. *Odyssey,* XIII, 109–112.
3. Guénon, op. cit., *Symbols of Sacred Science,* p160, n3.

Odysseus' return to Ithaca through the southern gate of the cave, the gate of the gods, is accommodated by Guénon's analysis. For Odysseus fulfils both of the conditions which are exceptions to the rule that human beings do not go in by the southern entrance. He is making a 'voluntary descent into the manifested world' insofar as he has deliberately foregone the unageing immortality offered by Calypso in order to return to Ithaca. But he is also 'a being representing the direct expression of a supra-cosmic principle' insofar as he is inalienably bound to the goddess Athene whose pre-eminence in wisdom among the gods he represents among people, as Athene tells him. On these grounds, then, we may compare Odysseus to a bodhisattva or an avatara. He is certainly no mere hero in the mould of Achilles or Hector. As Guénon indicates, such people are exceptional, and we remember that Odysseus was the last passenger on the Phaeacian service which closed down immediately afterwards. When Odysseus and Athene leave the cave, Athene places a stone over the entrance.

This interpretation must change our understanding of the very first line of the *Odyssey*, of the very first word, which is 'man'.

The man, O Muse, sing to me. . . .

For Odysseus is not quite a man. From the first moment when we meet him on Calypso's island he has already transcended the human condition: he has been offered ageless immortality. Nor should we be surprised that a poet who represents Athene as congratulating Odysseus on his lying tales, is himself given to verbal equivocation. If Odysseus is a man he cannot enter that southern entrance to the cave where men do not go in. If he does enter there he is not a man. But if Odysseus is not a man, then the *Odyssey is* a different kind of poem from the one we commonly think it. It is scripture. Odysseus is a superhuman being who voluntarily forsakes immortality in order to return to a mortal life. He returns to Ithaca, Penelope and Telemachus, as he risked his life to save his companions who had been trapped by Circe. His story, as he himself tells it, offers deliverance to everyone who hears it. The one epic in Greek literature which is named after a man is named after this man.

Why does Homer describe the cave as having entrances to the

north and south? Porphyry gives the reason for this in his essay on
the cave, attributing his interpretation to two of his predecessors,
Numenius and Cronius, who lived a century before him:

> Since, then, the present cave in an eminent degree is a symbol and
> image of the world, as Numenius and his familiar Cronius affirm, it is
> necessary in order to elucidate the reason of the position of the gates,
> to observe that there are two extremities in the heavens; viz. the win-
> ter-solstice, than which no part of heaven is nearer to the south; and
> the summer-solstice which is situated next to the north. But the sum-
> mer tropic ... is in Cancer, and the winter tropic in Capricorn.[1]

The northern entrance to the cave is the tropic of Cancer and the
southern entrance is the tropic of Capricorn. We should think of
the tropics of Cancer and Capricorn as times, not places. To us,
mention of the tropics of Cancer and Capricorn suggests rather the
two lines drawn round the globe parallel to the equator. But the
tropics are, more accurately, the moments when the sun changes
course, having reached the limits of its southward or northward
journey. If we take the interpretation in this way, then we need not
suppose that this Ithacan cave is symbolically transformed into the
whole earth and contains within itself the two tropic zones, while
remaining a speck on an island at about the thirty-eighth parallel
north. For while the earth as a whole is bounded by the lines of the
two tropics, every place on earth is similarly bounded, not spatially
but temporally. Though the two tropic lines do not run through
Ithaca, Ithaca is subject to the sun's movements which bring the
seasons and which determine the different characters of the days
which make the Ithacan year. Odysseus returns at the time of the
winter solstice, when the sun passes through the southern gate and
begins its northward course. In this way we can understand the
interpretation of Porphyry, Numenius, and Cronius as deriving
from a close reading of the text. They have noted the seasonal conditions at the time of Odysseus' return to Ithaca, and his remarks to
Eumaeus and Penelope about the end of the month and the year.

In the otherwise seamless continuum of time, the two solstices are

1. *Thomas Taylor the Platonist*, op. cit., Porphyry, p 309.

the first division, cutting the circle at diametrically opposite points. The circle is thus divided into two equal parts, the one the northward course of the sun, the other its southward course. The two solstices are appointed one to each of these two semicircles, so that each solstice serves as the gate or entrance to its respective half of the zodiac. The southern gate opens to the northward path, the northern gate to the southward path. The whole circuit of the year is regarded as the type or model of all the divisions of time, especially the month and the day, but also of a human life and of the life of a city. In the northern hemisphere the northward path of the sun occasions the period of growth, the southward path the period of decline. These two contrary processes of growth and decline are the necessary conditions of all coming into being and passing away. This synergy of growth and decline in the various allotments of time is here transposed into another dimension, so that it serves to represent the relation between the immortal and the mortal. In this passage the immortal corresponds to the ascending phase of growth, the mortal to the descending phase of decline.

If the two entrances to the cave are the southern and northern solstices, the southern solstice is the gate of the immortals and the northern solstice is the way down for people. The association of the southern solstice with the birth of an avatara in the northern hemisphere is well established. It is an integral part of Christian tradition, though not of Christian scripture. The corollary notion that the birth of everyone else is associated with the northern solstice in the northern hemisphere is not so well attested. There are, however, the group weddings at the Athenian festival of the Chalceia, when the sacred peplos of Athene was begun.[1] Nine months later the peplos was presented to Athene at the Panathenaea at midsummer, the New Year of the city, the birthday of the goddess and of many of her people.

* * *

So much for Porphyry's account of the Cave of the Nymphs. But we may still ask why Porphyry has chosen just this passage of the

1. K. Kerenyi, op. cit., p29.

Odyssey to give us the fullest symbolical interpretation which has survived to us from the ancient world. There is an answer to this question but it is not an answer which Porphyry himself offers at any point in his essay. We have seen how the northern and southern gates stand for the solstices and how the downward path of mortals passes from Cancer to Capricorn while the path of immortals is northward from Capricorn to Cancer. But these two directions northward and southward are the directions measured by latitude, by the Poles and polar zones, the tropics and the equator. It is just in these terms that Odysseus' travels from Troy to Ithaca are couched, according to Crates and Strabo. So may not this same figure of the sun's movements provide the framework of Odysseus' travels in general? The journey from Laestrygonia to the dead via Aeaea is the southward path; the journey from the dead to Ogygia via Aeaea and Thrinacie is the northward path. The southward path leads to mortal death, the northward path to immortality. Mortals are born from the north and perish into Hades; immortals return from the dead on their way to heaven. To be sure, the gates of the cave are the tropics while Odysseus' travels extend to the arctic and antarctic zones. But the symbolic meanings of northward and southward remain the same.

On this view, Odysseus returns from the land of his birth through the southern gate of the cave because ultimately he has come there from the dead who are southernmost. Figuratively, he is twice born; his passage through the southern entrance of the cave is his second birth, the initiatic re-enactment of birth common to many cultures as a symbol of spiritual rebirth. We may say that it is because Odysseus has visited the dead and returned that he is granted immortality, subject only to the condition that he successfully negotiate the horrors of Book XII. This chimes very well with the account of Heracles at the end of Book XI, the last of the ghosts with whom Odysseus talks before he returns to his ship. Heracles himself, we are told, is in heaven with trim ankled Youth. Only his ghost is in Hades. He commiserates with Odysseus on the harshness of mortal life and finishes by saying that he, too, was made to go down to Hades while still alive and bring back the three-headed dog, Cerberus.

Here is the connection between visiting Hades while still alive and a continued existence after death amongst the gods. The two lines describing Heracles in his post mortem bliss among the gods were often omitted by some ancient critics.[1] They were suspected of being an interpolation by Homer's Athenian editor, Onomacritus in the fifth century BC. Certainly, the post-mortem survival of Heracles as something more than a ghost seems alien to the beliefs expressed in the *Iliad* and *Odyssey*. But these lines fit perfectly into a reading of the *Odyssey* which takes Odysseus' travels as symbolic of the path to immortality. There is a second exception to the general beliefs about the afterlife in the *Iliad* and the *Odyssey*. Menelaus is told that he will not die in Argos but live on in the Elysian plain in the far west. The implication is that it is not fitting for Zeus' son-in-law to die.

It remains to ask whether Odysseus' route from the dead to Ithaca is the reverse of his route from Laestrygonia, to the dead. Do Odysseus' travels follow the plan of Heraclitus for whom 'The way up and the way down are the same' or is his return journey by a different route?[2] This matters because, in the case of the sun, the northward and southward journeys pass through different signs. The southing sun passes through Leo, Virgo, Libra, Scorpio and Sagittarius; the northing sun passes through Aquarius, Pisces, Aries, Taurus and Gemini. If Odysseus' travels are the sun's courses between the tropics projected onto all the latitudes between the Poles, then we would expect to find Odysseus on different paths as he comes to and goes from the dead. But in fact, he comes to the dead from Aeaea and goes back to Aeaea from the dead so that these portions of his northward and southward journeys are the same. On this ground, then, we may say with some confidence that Homer has not distinguished between the two paths except by direction. It is the same path travelled both ways and in this respect Odysseus' travels do not represent the sun's passage round the zodiac but only the sun's movements between the tropics.

But even in the case of the Aeaea-the dead-Aeaea stage of his journey, the matter is not quite simple. For when Odysseus returns

1. Footnote to Homer's *Odyssey*, XI, 602–604, tr. A.T. Murray (London: Loeb Classical Library, 1919), p428.
2. *Ancilla to the Presocratic Philosophers*, op. cit., Heraclitus, fr. 60.

to Aeaea from the dead, we are told something of Aeaea which we have not heard before during the long account of Odysseus' first visit there. This time, we are told that Aeaea is 'the dwelling of early dawn and her dancing lawns and the risings of the sun'. Dawn is to day what spring is to year, the mid point of the sun's ascending course. So the mention of this aspect of Aeaea is entirely appropriate to Odysseus' return to Aeaea where it would not have been appropriate to Odysseus' first visit. Then Aeaea would have been the dwelling and dancing place of dusk and the settings of the sun and in the west, not the east. And this is close to the description given of Odysseus' passage from Aeaea to the dead at the beginning of Book XI, which reaches its goal with the setting of the sun and the coming of night.

Apart from this slight oddity in the description of Aeaea, I see little reason at this stage for attempting to ascribe Odysseus' adventures to the zodiacal signs. Cancer and Capricorn are mentioned explicitly by Porphyry in his account of the Cave of the Nymphs, and the southern tropic and the South Pole are mentioned by Crates of Mallus according to Strabo. Beyond these terms we need not go. In any case the attribution of the adventures to the zodiacal signs is likely to create more difficulties than it resolves since the origin and meaning of these signs are wrapped in an even more impenetrable obscurity than are the adventures of Odysseus.

So ends the first part of this exegesis. Three distinct hermeneutical traditions have been explored: the astrogeographical or exoceanic tradition of Apollodorus, Strabo and Crates of Mallus; Plato's *Myth of Er*; and Porphyry's symbolical account of the Cave of the Nymphs. We have analyzed Odysseus' travels into two parts, a southward and a northward journey, whose turning point or tropic is the meeting with the dead within the antarctic circle. In the light of Plato's *Myth of Er* we have interpreted the adventures of Book XII as astronomical and interplanetary. Odysseus is qualified for Calypso's offer of immortality because he has returned from the dead and also because he has escaped the planetary whirlpool which conditions all mortal creatures. Passing north from Aeaea on the tropic of Capricorn, Odysseus crosses to the northern tropic through the vortex of the planets, sun and moon understood geocentrically. This should entail that Calypso and Ogygia mark the northern tropic, the tally to Circe and Aeaea. In that event we could identify Calypso's Ogygia as one of the West Indies just as we have located Circe off the coast of Queensland, Australia. But, alas, I have placed Phaeacia at the equator. If Calypso is on the northern tropic, then Odysseus goes south as well as west from Ogygia to Phaeacia and this disrupts the northward or ascending passage from the dead to Ithaca. At this point I must certainly conclude.

4

THE FALL
OF ODYSSEUS

IN THE FIRST PART OF THIS BOOK we have examined the astrogeographical, astronomical and calendrical framework within which
the *Odyssey* as a whole, and especially the travels of Odysseus, are
set. We have reached two general conclusions. Firstly the travels
of Odysseus were interpreted by some ancient commentators as taking him to all the latitudinal zones from the arctic to the antarctic,
and as taking him to the limits of the solar system. Secondly the
scientific reading is bound up in the minds of some of its ancient
exponents with a theory of spiritual regeneration, so that their
Homeric astronomy and geography are a map of the spiritual path.
The goal is nothing less than apotheosis, the turning of man into
god or immortal. Calypso offers Odysseus immortality and his
achievement is no less though he turns down Calypso's offer and
returns to Ithaca. On his return he appears to enter with Athene
through the southern gates of the Cave of the Nymphs, the gates of
the immortals.

It is time now to turn from these systemic and theoretical aspects
of the poem to a consideration of some of its episodes which we
have passed over till now or which we have considered independently of their stories: the Cyclops and Aeolus, Circe, Phaeacia and
the Battle in the Hall. As we enter the morass of interpretations to
which each one of these episodes gives rise we may hold on to and
direct ourselves by this hypothesis: that each of these episodes, by
virtue of its place in the general scheme which we have discussed,
represents or bears upon the theme of apotheosis. According to
Dante, a poem may be read at any or all of four levels of meaning:

the literal, the allegorical, the moral and the anagogical.[1] The first three names speak more or less for themselves but the fourth is unusual. It refers to that aspect of a poem by which it leads or draws us up to the divine. This, we may say, is the chief of the four levels of meaning and it is the one for which we will be searching in the *Odyssey.*

The study of Odysseus' travels in the *Odyssey* as symbols of sanctification may begin from Odysseus' original fall from grace, the moment of his alienation from the gods. This moment may be exactly located. It is when the Cyclops says near the end of Book IX:

> 'Hear me, Poseidon, Girdler of Earth, god of the sable locks. If I am yours indeed and you accept me as your son, grant that Odysseus, who styles himself Sacker of Cities and son of Laertes, may never reach his home in Ithaca. But if he is destined to reach his native land, to come once more to his own house and see his friends again, let him come late, in evil plight with all his comrades dead, and when he is landed by a foreign ship, let him find trouble in his home.' So Polyphemus prayed and the god of the sable locks heard his prayer.[2]

This first becomes binding at the exact moment when Poseidon, in whose name it is made, has heard it to the end and agreed to it. From that moment the next years of Odysseus and the lives of his crew are doomed. But it is not merely the blinding of the Cyclops which has brought this about. Odysseus, it seems, could have done all of that and still have escaped with no worse than the loss of six of his men. Odysseus provokes the curse by his reviling of the Cyclops and by his telling the Cyclops his name. Even having done all this, he might still have escaped because the Cyclops does not curse him immediately thereafter. Instead the Cyclops tries to lure him back to shore by offering the help of his father Poseidon and then realizes that his eye still has a chance of repair since his father Poseidon might heal it. But even now, having given his name and having been reminded that Polyphemus is the son of Poseidon, despite the warnings of the crew and after already having been

---

1. Dante, *The Convivio,* tr. R. Lansing, bk 2, ch. 1, details from 'Digital Dante' website.
2. *Odyssey,* IX, 528–535.

washed back to shore, while still in danger of that or of being smashed by a rock, Odysseus must still rant on and at last the curse is delivered! It is a comfort to know how very hard one has to work to be damned.

There is, of course, on our reading, a profound irony in this disaster which Odysseus brought upon himself. It is the same irony which governs the Christian reading of Adam's fall. If Odysseus had left behind the Cyclops and Hyperia without being cursed he would soon have reached Ithaca with his twelve ships and several hundred men. He would have settled down with Penelope after a ten-year absence, world famous for his exploits at Troy. Instead he is doomed to wander for many more years before he reaches home. Yet this extended journey is the very means of his salvation at a much deeper level, on our reading. His journey is the path of liberation and the Odysseus who finally returns to Ithaca through the gates of the immortals is utterly unlike the man who would have returned to Penelope all those years before. The Christian supposes the fall of Adam and Eve to be redeemed by the sacrifice of Christ, and not merely redeemed but utterly vindicated and justified, so that the whole sorry story of human history is shown to be contrived for the greater good and to have achieved it. In the same way the enmity of Poseidon becomes the key to spiritual salvation. In preventing Odysseus from returning to Ithaca, Poseidon enables his journey to immortality.

Why does Odysseus rant as he does in his last moments with Polyphemus? He is driven, it seems, by anger with a touch of pride. He may feel that the Cyclops has not paid adequately for the murder and eating of six of his men. He is anxious that Polyphemus knows his real name which suggests that his assumption of the false name No one has irked him. His fury and resentment are no doubt shared by his crew who were with him in the cave, but he does not speak for them. It seems to be a syndrome common to the Homeric heroes, this rush of blood to the head at the moment of completing some great feat of arms. They are filled with an overwhelming need to vaunt themselves and boast over their fallen antagonists. Odysseus tells the Cyclops how much he would like to have killed him, but even killing is not enough in some cases. Achilles swears he would like to cut up Hector's corpse and eat him raw.[1] Despite his master-ful leadership in the Cyclops' cave, Odysseus has lost six men and achieved nothing. Perhaps he feels the need to demonstrate to his companions how utterly he has dominated their common enemy, as a way of restoring his prestige. But however cunning Odysseus may be, he is less than fully rational here. He cannot know that the curse is coming but he knows full well the danger of Polyphemus' mis-siles. His rants are like explosions too long bottled up in his breast.

Odysseus' fatal outbursts in his last moments with the Cyclops are in marked contrast to the discretion and cunning which he has shown in the Cyclops' cave. There, once he has realized the danger, he scarcely makes a mistake. Consider his masterly reply to the Cyclops' request for information about his ship, his adoption of the name No one and his careful maintenance of his protests at the Cyclops' behavior even as he puts in train his plan for escape. It is a remarkable example of leadership and self discipline until the very last moment. Consider his silence and stillness as he hangs beneath the ram in the doorway while the Cyclops describes at length the murderous rage he feels against No one, and again the disciplined silence with which his companions from the cave rejoin their crew-mates and set out from shore. At this point a frown from Odysseus is enough to stop their weeping. Yet even so Odysseus almost man-ages to snatch defeat from the jaws of victory. He can no longer con-tain himself. In fact, though they all manage to escape physical violence at the Cyclops' hands, Odysseus has managed to snatch defeat from victory, since Poseidon's granting of the curse dooms each and every crewman to an early death as surely as if the Cyclops had caught them or smashed their ship. Off the island of Thrinacie their ship is at last destroyed by a missile from above, the thunder-bolt of Zeus which does all the work of the Cyclops. In just such a way his other ships are destroyed by the Laestrygonians. And all this just so that Odysseus can let off some steam. How the pressure must have been building inside him as each step in this terrifying, silent manoeuvre is accomplished!

1. *Iliad*, xxii, 346–347.

How can we be sure that this is the right way of reading this critical passage? Our reading receives confirmation of a kind from an
unexpected and authoritative source. First let us recapitulate three
elements in our account of the curse: it is a case of snatching defeat
from victory; their ship is returned to the shore as a result of the
Cyclops' first missile; it all comes about because Odysseus can no
longer restrain feelings to which he must give voice. These three elements may be found in the next story of the poem, the encounter
with Aeolus. The first two elements are obviously shared by the two
stories, the third less obviously. They snatch defeat from victory
when they open the bag of winds close enough to Ithaca to make
out individual human figures. As a result, the winds rush out of the
bag and blow them all the way back to Aeolia as the wash from the
Cyclops' first rock pushed them back to shore. And the bag of winds
itself is the breast of Odysseus, full of anger and pride, while the silver string is his vocal chords. The Aeolus episode is a repetition of
the last moment in the Cyclops episode, the inside story, as it were,
of Odysseus' fall. Certainly Odysseus is to blame in the Cyclops episode, while the companions are to blame for releasing the winds.
But in the Aeolus episode the ship itself represents Odysseus. The
sleeping Odysseus in the Aeolus episode represents the sleeping reason of Odysseus in the Cyclops episode.

On this view, the releasing of the bag of winds by the crew off the
coast of Ithaca corresponds to Odysseus' outbursts as he leaves the
Cyclops. The winds and the crew are the passionate feelings which he
can no longer contain. Accordingly the journey in the ship from
Aeolia to Ithaca corresponds to Odysseus' time in the cave down to the
boarding and setting out in their ship from the Cyclops' shore. During
all that time Odysseus is in perfect control of himself and his crew, and
they scarcely put a foot wrong. Even the lottery to choose Odysseus'
assistants in the blinding comes up with the very men Odysseus
himself would have chosen for the task. For ten days the ship sails from
Aeolia to Ithaca with a constant west wind behind them, the crew in
their places, Odysseus at the tiller, the other eleven ships beside them.
It is an image of perfect order, made absolutely secure by the fact that
all the winds which could blow them off course are safely stowed on
board where they can do no harm. This, I take it, is the image of a man

who has mastered himself, whose emotions are perfectly under control. But at the last moment, precisely because of his great success and within grasping distance of his goal, his concentration slips and all is undone. Everything he has achieved is reversed. Are the emotions of the companions when they open the bag the same as those which fuel Odysseus' tirade against the Cyclops? Anger and resentment are common to both, and above all the unholy glee in taking advantage when one's master or more powerful adversary is at a loss.

The bag of winds is the focus of the Aeolus episode and it is a symbol of containment and control. There are equivalent symbols in the description of Aeolia. The palace is surrounded by an unbroken wall of bronze atop cliffs which run sheer down to the sea. No explanation is given of how Odysseus manages to enter the palace. Aeolus and his wife have six sons and six daughters whom he has married to each other; and just in case we miss the thrill of incest here, the poet has taken care to describe the sleeping arrangements for these sturdy sons and their chaste wives, the blankets beneath them and the corded bed-frames beneath the blankets. All of this is counter to our intuitive notion of the warden of the winds. We would think the winds far reaching and dispersed and their warden likewise. Instead he seems to thrive on isolation, concentration and self-sufficiency. The only other description of Aeolia as an island is the one word 'floating', and this sounds rather more what we would expect of winds. This one word gives the clue to a quite different way of reading the Aeolus adventure, and by extension the story of Odysseus' fall. This way of reading is the last of Dante's four ways, the anagogical, which leads the student up to heaven. Compared to this, the interpretations I have given so far are merely moral. They make no claim upon the absolute. From here we must enter the fields of theology, metaphysics and ontology.

Imagine what there was before the universe began. I am not talking of chaos or even of a void. The state I wish you to imagine is without space or time. Neither of these has yet come into existence. This state, if state it can be called, was sometimes represented by the Goddess Night[1] or by Erebus. How from this did the universe arise?

---

1. *Ancilla to the Presocratic Philosophers*, op. cit., Parmenides, frag. 1.

It arose from a single point which existed nowhere, since there was no space for it to exist in, and at no time. And yet this first point contained within itself all space and time, and everything that was to be. And it began to expand:

> like the mass of a well rounded sphere stretching equally in all directions from the center.[1]

As it expanded it brought into existence space and time, since its expansion was simultaneously spatial and temporal, and the whole world came to be from what was within that first point. The bag of winds and the island of Aeolus with its walls and cliffs are both symbols of this point. The winds are the directions which spring out from the center, as well as the emotions which battle for expression in the heart. For the Greeks the four directions North, South, East, and West, received their names from the winds; for us the winds receive their names from their directions. The island of Aeolia, the winds in the bag, represent that state in which the whole world has turned back into its source and the being is inviolably centered in the absolute from which it arose.

This cosmogony has close parallels to our own biology. Each of us as living creatures sprang from seed which contained all the details of our natures. We too were once miniscule zygotes in our mothers' wombs and out of these has grown everything we have come to be. It is wise at least to entertain the notion that the whole world is a single living creature sprung from a single point. This cosmogony also has parallels with the theory of the big bang since in both the material universe begins from a single point. But in the big bang theories space already exists for the explosion to explode. This is also true of our growth; the space in which it occurs already exists. But it is not true of the world's seed. And where is that point now, where did it all start? If we ask this of the zygote from which each of our bodies has grown, we would find the question hard to answer. Since every single cell in our present body sprang from that zygote, we might say that the zygote is everywhere. But since every cell in our bodies sprang from the zygote, none of them can be the

---

1. *Ancilla to the Greek Philosophers,* Parmenides, frag. 8, line 30.

zygote, and so the answer must be that the zygote is nowhere. So the first point is both everywhere and nowhere, at all times and never. It is this that the floating of the island represents.

On the surface of the earth any point may serve to represent the center of the universe. For the Greeks the great center was Delphi, the womb from which the world was sprung. But they knew of others in other civilizations and they carried the symbolism into many aspects of their daily lives. The island of Delos was a floating island until Leto gave birth there to Apollo and Artemis. At this moment it was fixed to the bed of the sea. As Leto suffered labour she threw her arms around a palm tree,[1] the tree perhaps which Odysseus compares to Nausicaa.[2] Delos, too, marked the source and origin of things. In the Christian traditions every church is such a center. Cruciform churches with steeples or spires and crypts represent the six directions breaking out from the altar which is the very center, the one point through which one may return to the absolute. The churchgoer who approaches the altar is rolling up the world behind on the pathway to nonentity. The state of grace is to be in constant touch with that invisible root. This is Odysseus' state all the way from Aeolia to within sight of Ithaca, and when he leads so masterfully in the crisis of the Cyclops. This is the state from which he falls.

The pleasures of this state are amply enjoyed by Aeolus and his family. It is a Greek paradise: continuous feasting by day and at night the family repairs, wives with their husbands, to their springy beds. The wives are chaste, so there is no family discord. The hall is redolent of burning sacrifice, and rings with the sound of their feasting every day. Here we see the peculiar strengths of the anagogical as opposed to moral interpretation. On the moral reading of the Cyclops and Aeolus passages we are left with a strong negative precept: don't blow your top; bite your tongue; silence is golden. On the anagogical reading the message is positive. In the description of Aeolus' amenities we are given a glimpse of the beatific state. There is no sense here of thwarted energies as we might suppose with only

---

1. *Hesiod, The Homeric Hymns and Homerica,* op. cit., 'Homeric Hymn to Apollo on Delos', p117.
2. *Odyssey,* VI, 161–162.

the bag of winds for symbol. Instead we find a little city living in perfect equilibrium and unanimity, the like of which Odysseus does not find again until he reaches Phaeacia. By seeing the winds not merely as emotions but as directions, the story is made more general. It provides a symbol of cosmogenesis, in which the reading of the winds as emotions is only one part. It provides a symbol of heaven in the contented bliss of Aeolus and his family with boundless provisions and beyond all weather like the gods on Olympus. This is the company from which Odysseus is cast out.

Doomed by Poseidon, cast out by Aeolus, Odysseus suffers a twofold fall from grace. Without the Aeolus passage we would not see the inwardness of this. The Aeolus passage retells the story of the last moments of the encounter with the Cyclops, but in so different a way that there is no redundancy. The palace and manners of Aeolus and his family are the essential addition to the Cyclops story, to show us the divine or semi-divine state from which Odysseus falls.

We may compare the method here with the adventures of Book XII, three different stories which are all discernible within the single narrative of Plato's *Myth of Er* at the end of the *Republic*. Compare again what happens when we see that Scylla and Charybdis are the same, equally adequate symbols of the solar system presented now vertically, now horizontally. When the reader of Homer sees this identity for the first time, especially after knowing the story for a while, there is a thrill of recognition like that of solving a riddle. Newspapers once published photographs of familiar objects taken from unfamiliar angles and prizes would be awarded for successful recognition. Cryptic clues in crosswords sometimes describe familiar things in unfamiliar ways. Consider 'On the French Embassy staff'. If one takes 'on the staff' to mean part of the workforce, then the answer 'tricolor' will elude one even when all the cross letters have been filled in. But then at last the answer comes: the staff here is the flagstaff and we laugh with the setter for having duped us. When we finish, we survey all our answers for a final check and when we see 'tricolor', we smile again. That one we know we got right.

Of any event in the world there may be an indefinite number of likely accounts possible. Of any event in story there may well be a very large number, and certainly this is the case with these episodes in the *Odyssey.* How do we know which account is right, which Homer intended? This must have been as puzzling a question for his immediate audiences as for us now. In the sixth century BCE, Xenophanes sang that even if someone should happen to speak the truth in all its essentials, he could not know he had done so, since opinion is over all.[1] How do we know that we have understood Homer even if we have? This thrill of recognition is at once a pleasure by which the poet beguiles and amuses us, and a guarantee that we are following him. We know we have grasped his meaning when we smile at how stupid we were not to have grasped it before. From this point of view talking in riddles is an excellent way of making ones meaning clear. Jokes establish communication; the joker knows his audience has understood him when they laugh; and so do they.

Was Homer a riddler? If he was, on one late account he ended his life outriddled. He was told by the Delphic Oracle:

> The Isle of Ios is your mother's homeland and it will receive you when you die, but beware of the riddle of the children.

Homer eventually retired to Ios in his old age. Sitting by the sea one day, he asks some boys who were returning from fishing:

> Gentlemen, hunters of deep-sea prey, have we caught anything?

And they replied:

> All that we caught we left behind. What we did not catch we carry with us.

Homer did not understand this reply and asked them what they meant. They told him that they had caught nothing while fishing, but had spent the time hunting their body lice. Of these they had left behind the ones they caught, but still carried the ones which they had not caught. Then Homer remembered the oracle and prepared himself to die. Leaving the place he slipped on a muddy patch and fell on his side. Three days later he was dead. This story of Homer's death comes from a very late source, about the time of the Emperor Hadrian. But the riddle and Homer's failure to answer it

---

1. *Ancilla to the Presocratic Philosophers,* op. cit., Xenophanes, frag. 34.

are mentioned by Heraclitus in the sixth century BC when he claims that people are deceived by the most obvious things, as Homer was when he did not grasp this riddle though he was wiser than all the Greeks.[1]

There is little enough that we are told of Homer's life and this is an important part of it. We may have nothing here but the trope of wisdom abashed by childhood precocity. But the story of the riddle, especially in connection with the death, becomes much more pointed if Homer too, was considered a riddler, and this his humorous come-uppance. The retelling of an aspect of the Cyclops episode under the guise of the Aeolus episode, if this is what has happened, requires a lateral shift of mind which may not appeal to some. But it is often what is needed for piercing to the inwardness of myths and the appreciation of rituals and religious symbols. In this case it has one great advantage, that it is a game in which we have been given all the pieces and know all the moves. To see this relation between the Aeolus and Cyclops episodes we need just the text as we have it and nothing else. The text contains secretly within itself the solution to the riddle which it sets. It is its own explanation when we know how to look at it. This way of reading the two episodes provides a substantial explanation also for why the Aeolus episode follows immediately on the Cyclops episode. At this point the apparently random series of adventures resolves itself into the closest connection.

* * *

And still the question remains: was Odysseus justly punished for his outbursts against the Cyclops? What he did was not even a crime, at worst no more than an offence. We have examined how persistently Odysseus offends on this occasion, and we have reinterpreted the whole story in the light of the Aeolus passage. These considerations illuminate but they cannot justify what Odysseus suffers in consequence, and more importantly what his crew suffers. Because this one man speaks out of turn, hundreds of his companions are

---

1. *Ancilla to the Presocratic Philosophers,* op. cit., Heraclitus, frag. 56.

destroyed. Odysseus loses ten years of his life with Penelope, entirely misses the growing of his son and suffers great hardship, but on the principle that his is a happy fault which leads to the greater good of immortality, the price which Odysseus himself pays physically is not inapt. But for the companions, some of whom had tried to dissuade their leader from his reckless taunts, his actions meant death. And for Odysseus himself the most terrible part of the price which he paid was not his absence from Ithaca, wife and son, nor the hardship of his struggle to return: it was the knowledge that his foolishness had cost all his companions their lives. Were his own crew spared a little longer than the rest because they had tried to dissuade him? Did he see this knowledge in their eyes and the knowledge of their own doom because of him, these men with whom he had served the ten long years at Troy? With him they hear the Cyclops' curse and Teiresias' repetition of it among the dead. To be sure, Teiresias offers these companions some small hope. But this would hardly have assuaged Odysseus' guilt when they were all dead as Polyphemus had prayed they would be. The crews of the other eleven ships were given no chance at all. The man who washes up on Calypso's island, the laureate of immortality, is broken in body and mind, with all these dead companions on his conscience. He is like Oedipus at the ends of both of Sophocles' plays, or the violent, sad Heracles among the dead as he reflects on the agonies of life on earth.

How do we rate the lives of the companions? They are brought back to our minds at the end of the poem when Eupeithes, the father of Antinous, speaks of Odysseus to the Ithacan assembly

> 'Friends', he began, and tears for his son were streaming down his cheeks, 'I denounce Odysseus as the inveterate enemy of our race. Where is the gallant company he sailed away with? Lost by him, every one; and our good ships lost as well!'[1]

All this is true and the truth is worse than Eupeithes knows, for Odysseus himself was directly responsible for their loss and not merely a negligent or unfortunate commander. Was Odysseus unusually careless of his men? His relations with them are almost

---

1. *Odyssey*, XXIV, 426–428.

always cordial. He does not punish them for opening the bag of the winds. The nearest he comes to abusing them is when Eurylochus refuses to return to Circe's palace when the danger is over. At this point Odysseus considers killing him but decides instead to leave him behind. He is moved by the appearance of Elpenor among the dead and does what is asked of him by the youngest of the companions, who was in Odysseus' words 'not very strong in war nor in his wits'. On this occasion he pays close attention to the least of his charges. Odysseus also demonstrates his prowess in matters of morale in the *Iliad* when he reproves Thersites and makes his demagogic ranting look ridiculous to the troops. Thersites in his tirade against Agamemnon sounds very like the companions off the coast of Ithaca as they discuss the bag of winds. Compare the companions on Odysseus

> What a captain we have, welcomed wherever he goes and popular in every port! Back he comes from Troy with a splendid haul of plunder, though we who have gone every bit as far come home with empty hands....[1]

with Thersites to Agamemnon:

> Maybe you are short of gold, the ransom that some Trojan lord may come along with from the city, to free a son of his who has been tied up and brought in by myself or another of the men.[2]

With both Thersites and the companions Odysseus shows forbearance. His wits are so quick that he can easily isolate a trouble maker or a threat to his authority by bringing the other men over to his own side. There is not much sense here of that heroic ideal expressed by Heraclitus that one man is worth ten thousand if he is the best.[3] Odysseus is no autocrat. The real horror of Polyphemus' curse, for Odysseus more even than for the crew themselves, was that none of his companions would return but all would die. Odysseus was not only the unique survivor of a disaster which engulfed hundreds of his closest friends. He himself had precipitated the

---

1. *Odyssey*, x, 38–42.
2. *Iliad*, vi, 464–465.
3. *Ancilla to the Presocratic Philosophers*, op. cit., Heraclitus, frag. 49.

disaster. His psychological situation is exactly comparable to that of Coleridge's Ancient Mariner surrounded by the corpses of his ship-mates who fix with accusing eyes the survivor, the man who shot the albatross and doomed them but not himself. We need not agree that the punishment of Odysseus is justified to appreciate its effects and we can trace the damage which it did to him in everything which follows it. He has told the story to Calypso; he retells it in the court of Phaeacia; he tells it finally to Penelope. Like the Ancient Mariner,[1] Odysseus revives the trauma of that terrible journey to himself and in the minds and hearts of his hearers. It is his destiny henceforward to repeat it.

From the time he reaches Ogygia to the end of the book, proba-bly to the end of his life, Odysseus is a man of sorrows. He has hor-ribly transcended the man he once was. The work and the ambition of the hero now appall him. Thought of the years at Troy and the suffering of his journey provoke in him the most exquisite pain from which he cannot draw himself away. There has been in his heart an absolute breach with the heroic stereotype, and at the same time there is an irresistible fascination with the damage which that ideal has done to him. We can see all this in the words and actions of Odysseus at his first and second feast with the Phaeacians. His behavior is the exact reverse of what we would expect. Odysseus himself describes the scene for us when at last he begins to speak:

> Lord Alcinous, my most worshipful prince, it is indeed a lovely thing to hear a bard such as yours with a voice like the gods. I myself feel that there is nothing more delightful than when the festive mood reigns in a whole people's hearts and the banqueters listen to a min-strel from their seats in the hall, while the tables before them are laden with bread and meat, and a steward carries round the wine he has drawn from the bowl and fills their cups. This, to my way of thinking, is something very like perfection.[2]

Odysseus, though his name is as yet unknown to his host, is guest of honour at these feasts. His cup of happiness should be full. But

---

1. Samuel Taylor Coleridge, *Selected Poetry and Prose of Coleridge* (London: Random House, 1951), 'The Rime of the Ancient Mariner', 4, 430–441.

2. *Odyssey,* IX, 2–11.

there is much more to fill Odysseus' heart with joy than this. For the minstrel, the blind Demodocus, without knowing that Odysseus is present, begins the entertainment with a song about an argument between Achilles and Odysseus at another feast when Odysseus and Achilles served together under the walls of Troy. Here surely is the very acme of heroic satisfaction, to hear ones own deeds sung by a great minstrel at the very limits of the inhabited world, the last out-post of mankind. As he listens with the others, Odysseus begins to weep and covers his face with his cloak. Fond memories? Excess of joy? He covers his face because he is ashamed of weeping in front of the Phaeacians, and when the minstrel stops he uncovers his face and performs his social duties. But as soon as the minstrel starts, he weeps again. Alcinous alone sees what he is doing and hears him groaning because they are sitting next to each other. At this point, very tactfully, Alcinous suggests that the company has enjoyed the food and music and should move to other entertainments.

That evening things get worse. This time it is Odysseus himself, still anonymous, who asks Demodocus for a song about Odysseus, the story of the Trojan horse:

> Odysseus broke down as the famous minstrel sang this lay and his cheeks were wet from the tears that ran from his eyes. He wept as a woman weeps when she throws her arms around the body of her beloved husband, fallen in battle before his city and his comrades, fighting to save his home town and his children from disaster. She has found him gasping in the throes of death; she clings to him and lifts her voice in lamentation. But the enemy come up and belabor her back and shoulders with spears, as they lead her off into slavery and a life of miserable toil, with her cheeks wasted by her pitiful grief. Equally pitiful were the tears that welled up in Odysseus' eyes, and though he succeeded in hiding them from all but the king, Alcinous could not help observing his condition; he was sitting next to him and heard his heavy groans.[1]

This time there is no help for it. Alcinous speaks out over the minstrel and asks Demodocus to stop playing, for he does not please all with his song. And then, brilliantly, Alcinous launches into his

---

1. *Odyssey,* VIII, 526–534.

longest speech, some fifty lines, as he presses Odysseus to tell them who he is. The shocking breach in the fabric of the feast is instantly repaired but an enormous pressure is put on Odysseus to speak his name. The whole assembly waits upon him in the silence which follows, the minstrel as expectant as the rest. Alcinous finishes his long speech on a most loving note: 'For a sympathetic friend can be quite as dear as a brother'.

And so at last Odysseus yields and tells them his name after praising the feast. But it is with the precise quality of his grief that Homer is concerned in the long passage just quoted. Just when we would expect Odysseus at the zenith of his self esteem he is in fact plunged to the very nadir of grief. He is feeling what Hector prayed he would never hear Andromache express:

> Ah, may the earth lie deep on my dead body before I hear the screams you utter as they drag you off! [1]

But here the feeling is intensified by the woman's being dragged from the body of her husband in the very last moments of his life. How does Odysseus come to be feeling this? Demodocus is just describing Odysseus' heroic fight to the death with Deiphobus in the palace on the night of Troy's sack. Far from being like the woman in the simile, Odysseus might be the man who laid her husband low as he has just felled Deiphobus, or he might be the officer commanding the troops who are beating her on the back and shoulders with their spears. Instead he is the woman herself and relives the sack of Troy as if he were one of his own victims. Yeats suggested that when we die we experience everything done to us as if we did it and everything we have done as if it were done to us. [2] Odysseus has learnt the meaning of his own fierce doings to the last bitter tear.

But Odysseus himself asks for the song of Troy's sack. He has already suffered on hearing the song of his argument with Achilles. Why now does he deliberately risk further torment by his request?

1. *Iliad*, VI, 464–465.
2. William Butler Yeats, *A Vision* (London: MacMillan and Company, 1962), p 238.

He seems to be flagellating himself like Coleridge's mariner, and we may wonder at this point whether the mental disturbance which is manifest in such a self induced paroxysm of guilt will lead to sainthood or to massacre. In the event it leads to both. In Virgil's account of Troy's sack Odysseus savagely mutilates Deiphobus' body after killing him.[1] It may be that Virgil was seeking to supply another reason for Odysseus' collapse at the Phaeacian feast. On the strength of what Homer gives us in the *Odyssey* Odysseus is at once fascinated and revolted by certain passages in his own past.

How does Odysseus' emotional state at this point in his life compare with what we know from Homer of the other heroes after Troy? Nestor has readjusted perfectly. A much older man when he went to Troy, he seems to have accommodated to its rigors and to the ending of them. Menelaus and Helen are settled but are still on medication for depression. Ajax and Agamemnon are dead but continue to nurse their resentments in Hades. Perhaps Achilles in Hades is closest to the weeping Odysseus here. Achilles, too, has transcended the heroic ideals which bound him on earth. He had laid waste to others' lives until the river was choked with corpses and itself rebelled against him,[2] and he had chosen a short life with glory rather than a long one without it.[3] But now he would rather be the slave of a man without land, as long as he was on earth, than king of all the dead. This is measurable with Odysseus' state. And when Odysseus first heard Demodocus sing of the argument between Odysseus and Achilles during the siege of Troy, his tears and groans might have sprung from remembering these words of Achilles among the dead. As he heard again, on Demodocus' lips, the words of the living Achilles, he would compare them with those of Achilles in Hades. All that pride and vigor would sound like the emptiest bombast or worse. And so, too, would his own words in reply to Achilles.

Odysseus may be very like Achilles in another way, not now the

1. Virgil, *Aeneid*, tr. W.F. Jackson-Knight (Middlesex: Penguin Books, Ltd., 1958), VI. 528.

2. *Iliad*, XXI, 218–221.

3. Ibid., IX, 410–416.

Achilles in Hades but the Achilles after Patroclus' death. Driven by grief and remorse at Patroclus' death, Achilles is deranged. With the too scrupulous honour of a soldier, Achilles believes himself responsible for his friend's death, and this triggers a berserker rage, a homicidal mania, in which the killing by the river is a kind of mindlessness. We can see how absolute that mindlessness becomes in the film *Lawrence of Arabia*, when Lawrence and the Arabs massacre the Turkish troops who had attacked an Arab village. Odysseus, too, is deranged by grief and remorse for the death of his companions, but his madness, if it is madness, takes him differently. Instead of the berserker rage, the stalk. But the end is the same, massacre, and a homicidal mania might as well characterize Odysseus' killings as Achilles'. On this view, the deaths of the suitors pay for the deaths of the companions as though Odysseus somehow identified the suitors as the killers of the companions. But I do not think Odysseus is mad: it is to the Achilles in Hades that we should compare Odysseus after Ogygia, not to the mad Achilles in the second half of the *Iliad*. But the comparison with the mad Achilles suggests the quality of suffering which has broken Odysseus from his former habits of thought and feeling. Achilles lost only Patroclus; Odysseus lost everyone. When Odysseus finally gives Alcinous and the Phaeacians his name, in what tone exactly does he say the next words?

> I am Odysseus, Laertes' son. The whole world talks of my stratagems, my fame has reached the heavens.[1]

I can hear no satisfaction here, no pride. He is just that Odysseus of whom they have been speaking, the man who devised the horse, to whose fame Demodocus and the Phaeacians have themselves attested.

Odysseus is not a homicidal maniac in the battle in the hall, however excessive his actions may seem. At no point does he lose command of himself. This is not the mindlessness of Achilles at the river or of Lawrence massacring the Turks, though like them Odysseus is covered with blood. He exercises acute discrimination in the rapid judgements he makes of suppliants during the battle, of Leodes,

---

1. *Odyssey,* IX, 19–20.

Phemius and Medon. As a beggar in the hall he has had an opportunity to assess them all. He even smiles at Medon, to allay his anxiety, when he spares him. Finally his words to Eurycleia suggest a mind not merely sane, but more sane than almost any other which we have experienced in the *Iliad* and the *Odyssey*. Eurycleia comes for the first time upon the slaughter and the bloody Odysseus in its midst, and she is on the point of shouting her joy when Odysseus checks her:

> But when Eurycleia saw the dead men and that sea of blood her instinct was to raise a yell of triumph at the mighty achievement that confronted her. Odysseus, however, checked her exuberance with a sharp rebuke: 'Restrain yourself, old dame, and gloat in silence. I'll have no jubilation here. It is an impious thing to exult over the slain. These men fell victims to the hand of heaven and their own infamy. They paid respect to no one who came near them - good men and bad were all alike to them. And now their own insensate wickedness has brought them to this awful end.[1]

Perhaps Odysseus feels here that he has not performed a heroic deed, has not met men face to face in the battle line or single combat. He has been destroying dangerous but base vermin. But it is his view now that it is impious to exult over dead men, and in any case it is not he but the gods who have destroyed them, the gods and their own wicked deeds. All his planning, his endurance of humiliation, his courage and his extraordinary fighting skills are not his any more, but the agency of a divine power for justice. He has far more to boast of now than after blinding and escaping from the Cyclops, but he has learned that lesson so that not even his present success can shake it.

There are two moments in Homer's work when a character steps off the page and is unmistakably in the room with me. This speech to Eurycleia is one of them. The other is in the ninth book of the *Iliad* when Achilles says to Odysseus and others:

> For life, as I see it, is not to be set off either against the fabled wealth of splendid Ilium in the peaceful days before the Achaeans came, or against all the treasure that is piled up in rocky Pytho behind the

1. *Odyssey*, XXII, 411–416.

marble threshold of the archer king Apollo. Cattle and sturdy sheep can be had for the taking, and tripods and chestnut horses can be bought, but you cannot steal or buy back a man's life when once the breath has left his lips.[1]

These two are the moments they are because in each is spoken my own heart's thought about the world these books describe. How difficult it is to hold ones mind and feeling to a fictional world full of violent men at their most violent! And then in the very midst of it, the most violent of them all speak purest sense about their violence. Achilles does not listen to his own words but stays in the war and dies. When we meet Achilles again in Hades, he returns to just this theme but now, too late, he does know the truth of his words. Odysseus knows the truth he speaks to Eurycleia, and it is not too late. His knowing it brings to an end the cycle that began with the curse of Polyphemus. At the most terrible cost Odysseus has come to see his error.

In the same way the battle in the hall and the speech to Eurycleia bring to an end the Homeric epic itself. The Trojan cycle concludes here with the settling of the destiny of the last hero to return. The view we are given of this is entirely positive, however terrible the carnage. It is a restoration of order after chaos. Simone Weil supposes the *Iliad* to be the epic of force. The story of the *Odyssey*, too, is the story of force which usurps the proper authority of Odysseus' kingdom and palace. When this proper authority is re-established, the disorder which has been the poet's subject for forty-five books is finally over. In this disorder the exultation of the victor over the defeated has played a continuous part. We remember Patroclus' crowing over the dive of Cebriones; Hector's boasting over Patroclus; Achilles' over Hector.[2] At their moments of triumph these heroes, like Odysseus with the Cyclops, are carried on to arrogance, and commit themselves to their own deaths by an unpardonable boast. No wonder that in their triumphal procession the Romans made a slave repeat in the ear of the victorious general 'Remember: you must die.' It is just such a wisdom which Odysseus displays to

1. *Iliad*, IX, 401–405.
2. *Iliad*, XVI, 745–750; XVI, 830–842; XXII, 331–336.

Eurycleia. The New Testament transcends the Old Testament when it displaces the *Lex Talionis*, an eye for an eye and a tooth for a tooth, with the principle: resist not evil. In a similar way the *Odyssey* transcends the *Iliad*. Like the *Odyssey* the New Testament may be read by itself. But again like the *Odyssey* it gains immeasurably in moral force when it is read in the context of the earlier story. The *Iliad* and the *Odyssey* comprise a single, complex moral which may be felt most acutely in the *Odyssey*'s treatment of Odysseus' relations to Achilles and at its end.

Before the battle in the hall, Odysseus strings his bow and shoots his arrow through the axeheads. What this means as a symbol of the sun's yearly course through the months has been discussed. As a symbol of Odysseus' own passage through the trials of his *Odyssey* it is comparable to the eye of a needle through which it is easier for a camel to pass than it is for a rich man to enter the kingdom of heaven. But it is also a symbol representing the most rigorous exactitude. The same niceness of judgement which allows Odysseus to shoot that arrow through all twelve axeheads is, we should infer, shown by his judgement and execution of that judgement on the suitors. Again, accurate shooting requires equanimity and Odysseus is about to provoke the greatest and most dangerous battle of his life. The calmness of his shooting is extraordinary. When he has just strung the bow Odysseus tries it and the string sounds like the note of a swallow. This pure note shows that the bow is sound and true, and it shows that the man is sound and true. It is the overture to the dance and song of the battle and we may imagine that the minstrel here would loudly pluck a single string and let the note resound. This is a musical symbol of a new purity in the affairs of men.

But if Odysseus is the same before, during and after the battle in the hall, then the breakdown in the court of Phaeacia is an act of contrition. Even his asking for the song which triggers the breakdown is an act of contrition rather than a symptom of war psychosis. In that case we must read his behavior in his palace before the battle a little differently too. He is not merely stalking his prey, nor putting them under surveillance to determine whether they should be spared or not. Now as beggar in his own palace, Odysseus is not the man he was when he beat himself and dressed in rags to pass as a slave in the streets of Troy during the war. That was no penance but great courage and cunning. But after weeping at Demodocus' lay, the blows and insults which he suffers at the hands of the suitors are strangely a requital for his own pride and its consequences to others. He was humble, too, in the Cyclops' cave when he needed to be. He bit his tongue and served the Cyclops wine, and perhaps because of that forced humility he could not restrain himself when the chance came to speak his mind. In the hall the provocations are more personal and more intense and he bites his tongue again and again. But when at last he is at leisure to exult, no word of it escapes him. The anger which he feels at the behavior of the suitors is neither expressed nor repressed, for if it were repressed it would break out at last. Instead that anger is, I suggest, constantly discharged through the same remorse which he shows in Phaeacia. In this way his return to his palace is a penitential pilgrimage, a kind of passion in the Christian sense. This penance does not end with his restoration in the palace. It is not a seasonal adoption of the lowest role, like that of the paterfamilias in the Saturnalia. For Teiresias has told him that he has yet to go on that last mission for Poseidon, with his oar over his shoulder, to find people who know nothing of the sea or ships. We have no song of this last mission but we do have a minute account of the bitter humiliations which Odysseus as a beggar receives at the suitors' hands. He suffers them as he does because he knows that he deserves them, just as he will leave Penelope again for that last mission, and for whatever else may be required of him.

And still the question remains: were Odysseus and his companions justly punished for his outbursts against the Cyclops? Odysseus' behavior here is characteristic of the hero, and Odysseus' sufferings as a result purge him of those values and turn him into a different kind of man. No less a mighty warrior, he is broken somewhere in himself. Does this justify the destruction of all his companions? Are the gods in general and Poseidon right to destroy all these men as a way of punishing and purifying Odysseus? The answer may be that visiting the consequences of a commander's error on his troops is simply how things are. It is the price of authority that the led may die because the leader fails and that the leader must then bear the weight of this knowledge. In the case of a sensitive officer it is hard to know whether this is a fate worse than death. Achilles and Odysseus are both sensitive officers who snap under the strain of sending men out to die. However much sympathy we feel for the lost companions, the poem and its hero feel more. Grief and remorse rack Odysseus, the one flaw in his indomitable, heroic will. To insult and abuse the Cyclops does not count for much in the calendar of violence which is the *Iliad* and *Odyssey*. Not to insult the Cyclops, not to boast over the suitors, this counts for everything. For the man who has learned not to do this has not merely learned to bite his tongue, but he has been broken in pieces and reconfigured. The stronger the will, the greater the pressure needed to break it. The death of all the companions is pressure enough.

Once in a class Swami Chinmayananda was asked the following question: 'why, Swamiji, if you are liberated from your mortal shell and firmly established in the absolute, are you still here?' This is a little like the question: why does Odysseus turn down Calypso's offer of immortality on Ogygia to return to Ithaca, if immortality is the highest goal of mortal existence? The Swami replied by asking the class to imagine an electric fan suspended from the ceiling. Its switch was located by the door. When the switch was on, the current flowed to the fan's motor and the blades turned. But when the fan was switched off, the blades continued to turn for a while even though there was no current flowing to the motor. In the same way the person liberated while still alive continues to act in the world even though the motive energy for doing so has ceased. We were, I think sceptical of this answer because we knew that though the Swami was diabetic, had had a multiple heart bypass and was in his mid seventies, he still worked twenty hours every day. But the answer is suggestive of Odysseus' state from Ogygia onwards. Even under extreme provocation, as with Euryalus in Phaeacia or the suitors, Odysseus is hard to stir. When he acts he acts with dispassion. Though he suffers he is not fearful nor anxious that he will fail. He sees himself as an agent of the gods and his presence is surrounded by portents of doom for the suitors and of victory for himself. He has become uncanny.

5

THE BELLY
OF THE CYCLOPS

The Cyclopes, including Polyphemus, are human beings. They suffer, if that is the word, from congenital monocularity and gigantism, but they are still people. They are rather hard to imagine. Homer does not tell us how tall Polyphemus is, only that his staff is like the mast of a merchantman, and his strength is more than that of twenty-two four wheeled wagons when he moves his rock door. Again it is hard to visualize how the human face would look with a single eye, especially as the Cyclops has brows. Could the human frame bear a weight and height so much greater than normal? The variations between the largest and smallest fish or between the largest and smallest spiders are, it is true, greater than that between the Cyclops and Odysseus, but in the first case the creatures are borne up by water and in the second by four times the number of legs. This, I dare say, is a consideration which Homer overlooked. But his inclusivity is commendable.

We do gain some sense of Polyphemus' weight, not through measurement nor even through his physical actions but through the slowness of his wits. But his wits are all there. He can make a joke of the kind which sadists make when he rewards Odysseus for the wine by telling him that his guest gift to Odysseus in return is that Odysseus will die last of all his companions. To call such a postponement a guest gift is heavily ironic as though he wishes to turn Odysseus' expectations of hospitality against him. He seems particularly resentful of Odysseus' lecturing him on his duties as a host. He may even be eating Odysseus' men just to make the point to Odysseus that he is not bound by such duties but goes his own sweet way. The Cyclops is also capable of deceit, though of a very unconvincing kind, when he tries to entice Odysseus back to land with promises of entertainment and the help of his father Poseidon. The Cyclops may even be musical. He certainly whistles on leaving the cave in the morning, though this may simply be his way of controlling his flocks. It is a happy picture, the giant strolling off in the morning, his heart gladdened by the thought of the tasty human morsels awaiting his homecoming, safe and fresh in that cave of his.

Odysseus irritates Polyphemus by his insistence on the giant's duties as a host. Polyphemus refuses to acknowledge these duties. He claims that the Cyclopes pay no attention to Zeus or the gods because the Cyclopes are stronger than the gods. Whatever the reason, Polyphemus certainly appears unmindful of his father Poseidon when he is blinded. In the middle of the night the other Cyclopes have told him to keep quiet and pray to his father Poseidon if he has a problem, but their advice it seems, did not sink in. It is not until Polyphemus tries to lure Odysseus back with the unconvincing promise of entertainment and the help of his father Poseidon that Polyphemus himself seems finally to realize that his father might heal him. It is as though his own mention of Poseidon's name reminds Polyphemus that he can seek a cure for his blindness in that quarter. The realization comes as a shock and for a moment he forgets his conversation with Odysseus to muse on it. Alcinous later tells Odysseus that the Cyclopes like the Phaeacians and the wild tribes of the giants are close to the gods. But we see that where the Phaeacians are devout, the Cyclopes are blasphemous and sacriligious. Polyphemus himself would, no doubt, put this difference down to their different sizes. The Cyclopes do not need to respect the gods and so they do not. So it is strange to see Polyphemus praying to Poseidon at the end of this episode. In his extremity even this bullying individualist turns to a power which he now admits to be without equal among gods and men.

Polyphemus feels no reverence for the gods when Odysseus first meets him, so he does not feel himself bound by divine law. Again the Cyclopes in general have nothing to do with human laws or assemblies though they live within shouting distance of each other. They must associate in some ways if only to reproduce, and we learn that Polyphemus has had his fortune told by Telemus, a Cyclops gifted in the art. But on the whole the Cyclopes live apart from each other in caves on mountain peaks, and each of them gives his own laws to his wives and children. It is lucky that Odysseus and his companions lit upon the cave of a bachelor. They would have been hard put to cope with a family of the monsters. Polyphemus' solitariness is, nevertheless characteristically Cyclopean, and Homer tells us that he was the strongest of them all. He is also a fine shepherd. However chaotic his relations with gods and people, his management of his flocks is impeccable, a model of order and regularity. His intimate knowledge of his animals' ways makes him question his favorite ram when the ram is last to leave the cave rather than first. It is in these animals, it seems, that Polyphemus has invested his humanity. For people he cares nothing. His first questions for Odysseus who is sitting quietly in the cave with his companions reveal his suspicious nature. He suspects them of piracy, though he has no reason to do so. On the other hand it is true that Odysseus and his men did raid the Cicones in just the way that Polyphemus suspects. And Odysseus at least is armed with a sword.

What do the Cyclopes normally eat? We are told that without planting or tilling, wheat, barley and vines spring up and from the vines the Cyclopes make wine. In Polyphemus' cave there are cheeses and whey, and Polyphemus is going to make his supper from the milk which he has just drawn from his flocks. But there is no suggestion in the story that he eats these animals, nor that he uses fire for the cooking of meat, the baking of bread or even the heating of gruel. Apart from Odysseus' companions, his diet appears to be vegetarian, though not vegan. The Cyclopes need to do very little to live well. Perhaps the way the cereals and the vines grow without planting or ploughing explains the enormous size of the Cyclopes. This fecundity takes away the least justification for the eating of Odysseus' men for there is other food available to Polyphemus with as little effort as he expends on the companions whom he eats. He eats these men raw without even taking their clothes off and he keeps the others fresh by keeping them alive in his cave until just before he eats them. Then he tears or cuts them limb from limb:

[He] jumped up, and reaching out towards my men, seized a couple and dashed their heads against the floor as though they had been puppies. Their brains ran out on the ground and soaked the earth. Limb by limb he tore them to pieces to make his meal, which he devoured like a mountain lion, never pausing until entrails and flesh, marrow and bones, were all consumed, while we could do nothing but weep and lift up our hands to Zeus in horror at the ghastly sight, paralyzed by our sense of utter helplessness. When the Cyclops had filled his great belly with this meal of human flesh, which he washed down with unwatered milk, he stretched himself out for sleep among his flocks inside his cave.[1]

From within the story we may explain this extraordinary behavior by supposing that Polyphemus has somehow acquired a taste for human flesh or that he has been infuriated by Odysseus' whining about the gods and his duties to strangers. For narrative force this is one of the very highest points in the *Odyssey*. The image of the cannibal at this point sears the mind; it is this image which more than any other makes the Cyclops story unforgettable. But does this ugly and terrible story serve a religious or spiritual purpose? Why does the Muse inspire it? The Cyclops episode is arguably the most accessible of any passage in Homer's epics to an anagogical or spiritual reading.

Why does Homer represent Odysseus as trapped in the cave of a giant man who devours his companions and means to devour Odysseus himself? Because this neatly turns the tables on human beings as carnivores. Here Odysseus suffers at the hands of a man exactly what all those goats and pigs, oxen and deer and sheep have suffered at his hands when he ate them. We have here a reversal very similar to the one which we have examined in Alcinous' court. When Odysseus breaks down and weeps like a woman at Demodocus' description of his warrior prowess at the sack of Troy, he is experiencing as victim what he inflicted as victor. In the Cyclops' cave it is the same. Here, too, Odysseus is made to feel the violence of his own fierce doings. As Homer tells the story, Odysseus' weeping in Phaeacia comes before the Cyclops episode by a few hundred

---

1. *Odyssey,* IX, 288–298.

THE BELLY OF THE CYCLOPS 107

lines. As the epic unfolds, Odysseus' breakdown in Phaeacia alerts us to the possibility of a similar reversal in the Cyclops episode.

Eating is ugly even without the Cyclops' lack of manners. Worse than ugly, it is degrading. Later Greek thinkers were especially conscious of this. Incarnation as a mortal creature commits the fallen soul to the incessant struggle for the means to subsist in a war of all against all. In the case of human beings the extremity of this degradation is cannibalism; next comes the eating of meat; and last comes vegetarianism. If it were possible, Porphyry says in his essay on abstinence from animal food, we should abstain from all food.[1] In every case the living human survives by the tearing to pieces and grinding to chyme of other living creatures. The whole world is this appalling feasting of creature upon creature, tearing at the flesh with talon and tooth. Justice, according to Hesiod, begins when members of a class or species do not eat each other:

> For the son of Cronos has ordained this law for men, that fishes and beasts and winged fowl should devour one another, for Justice is not in them; but to mankind he gave Justice which proves far the best.[2]

People are just as a species because they do not eat each other, but only the gods are really just. They eat no creatures at all but only nectar and ambrosia. That is the ideal condition from which we have been precipitated into mortal life, to eat and to be eaten. And every mortal creature is in this fix. Even the dead thirst for blood. The companions are not cannibals, though some of them are eaten by a cannibal. But this does not make that requital excessive. The companions are subjected to exactly what they have inflicted on all those other creatures: to be eaten by a man.

The cave of the Cyclops is a symbol of the human belly. It is a very literal symbol: the floor of the cave is covered by animal dung, enough of it to make the hiding of the six-foot section of staff easy. The Cyclops lights and heats his cave with a wood fire which must aid decomposition and help to release the stink. Scattered here and

---

1. Porphyry, *On Abstinence from Animal Food*, tr. Thomas Taylor (London: Centaur Press, 1965), p 46.

2. *Hesiod, The Homeric Hymns and Homerica*, op. cit., 'Works and Days', pp 22–25.

there are the six little pools of human brains where the Cyclops has smashed the skulls of the six companions. When Odysseus makes the Cyclops drunk, the giant falls over. Fortunately for him his great neck is turned or he would have choked on his own vomit, the mess of partly digested humans and strong wine which pours from his mouth as he sleeps. The stench alone must have been stupefying. This cave each of us carries around inside. We can never escape from it until we cease to exist as creatures. Possibly these details of the Cyclops story work even more powerfully on our twenty-first century imaginations than they did on the ancient Greeks. Our sanitation has gone to great lengths to preserve our awareness from these disgusting facts.

The most powerful exponent of the horrors of eating after Homer is the fifth century philosopher-poet Empedocles. In his poem *Purifications* Empedocles denounces the eating of living creatures as the greatest pollution of mankind. His abhorrence is driven by a sense of how the eater and the eaten are interrelated:

> Will ye not cease from this harsh sounding slaughter? Do you not see that you are devouring one another in the thoughtlessness of your minds?[1]

This corresponds very closely to Homer's choice of a man-giant to eat Odysseus' companions. But Empedocles outdoes Homer's horror when he introduces the closest familial relations as binding victor and victims:

> The father having lifted up the son slaughters him with a prayer, in his great folly. But they are troubled at sacrificing one who begs for mercy. But he, on the other hand, deaf to the victim's cries, slaughters him in his halls and prepares the evil feast. Likewise son takes father and children their mother, and tearing out the life, eat the flesh of their own kin.[2]

Empedocles believed that he himself once lived as a plant.[3] It is

---

1. *Ancilla to the Presocratic Philosophers,* op. cit., Empedocles, frag. 136.
2. Ibid., frag. 137.
3. Ibid.

very hard to see how any living creature could avoid this kind of crime, whatever it ate.

The filth of the cave does not afflict Odysseus and his companions. It is the killing and tearing of the flesh which horrifies them. As they watch the Cyclops killing and eating the first two of their comrades, they weep and lift up their hands to Zeus spontaneously, it seems, or even automatically as though compelled to the gesture by the horrible vision before them and by their own helplessness. Horror at eating is not confined in the Greek tradition to acts of cannibalism but extends to meat eating in general and perhaps even to the eating of beans. The Pythagorean prohibition on the eating of beans, if such there was, is explained by some authors as arising from the similarity between the shape of the bean and the shape of the human embryo. The Pythagoreans were vegetarians for the most part and believed in the kinship of all life since human souls could transmigrate into the bodies of animals. Such a notion underlies Homer's account of how the companions were turned into pigs by Circe; and we must entertain the likelihood that the stag which Odysseus kills and which they all eat on Circe's island was once a human being before meeting Circe. It is generally supposed by modern scholars that this revulsion from meat eating was post Homeric and belongs to the sixth century, to Pythagorean and Orphic teaching. But there is nothing in extant Greek literature to top the power of Homer's description here. Orphism and vegetarianism, in their attitudes to the eating of animals, may as easily be derived from these Homeric teachings as original to post Homeric Greece.

This, then is the condition of all mortal creatures, to be condemned to killing and eating their own kinsfolk so that they may themselves survive. In this respect the Greek sensibility seems to differ from the Jewish. In the story of Adam and Eve it is the eating of the fruit of a particular tree which ensures their downfall; in the Greek tradition it is the act itself of eating. It is true that the companions are doomed because they eat the Cattle of the Sun and not any cattle; and that they are turned into pigs because they eat Circe's posset and no other. But in the Hymn to Demeter Persephone is condemned ever to return to Hades because she eats some

pomegranate seeds, and there is nothing special about these seeds.[1]
Consider the following lines of Empedocles:

> There is an oracle of Necessity, an ancient decree of the gods, eternal,
> sealed fast with broad oaths, that when one of the divine spirits
> whose portion is long life sinfully stains his own limbs with blood-
> shed, and following Hate has sworn a false oath - these must wander
> for thrice ten thousand seasons far from the company of the blessed,
> being born throughout the period into all kinds of mortal shapes,
> which exchange one hard way of life for another. For the mighty Air
> chases them into the Sea, and the Sea spews them forth onto the dry
> land, and the Earth towards the rays of the blazing Sun; and the Sun
> hurls them into the eddies of Aether. One receives them from the
> other, and all loathe them. Of this number am I too now, a fugitive
> from heaven and a wanderer, because I trusted in raging Hate.[2]

On this view we are all fallen divine spirits who have stained our
limbs with bloodshed. This is related to the fragments of Empedocles which I quoted above. This sinful staining occurs when we
tear to pieces our kinsfolk. But the parts of our human bodies, our
limbs, are largely formed of blood themselves. Our being human
requires our being insanguinated. So this staining of our limbs with
bloodshed has the twofold meaning of our draining the lives of
other creatures to nourish ourselves and of our being largely composed of blood as a result.

Homer and Orphics such as Empedocles share a deep concern
over the killing and eating of living creatures by living creatures.
The Orphics believed in the transmigration of the souls from
humans to other species and such a belief appears in a very limited
form in Homer's account of Circe's spells. We do not find in Homer
any belief in the superiority of the vegetarian diet over the carnivo-
rous. Nor do we find in Homer the disgust at blood sacrifice we find
elsewhere as early as Heraclitus:

> They purify themselves by staining themselves with other blood as if
> one were to step into mud to wash off mud. But a man would be

1. *Hesiod, The Homeric Hymns and Homerica*, op. cit., 'Hymn to Demeter'.
2. *Ancilla to the Presocratic Philosophers*, op. cit., Empedocles, frag. 115.

thought mad if any of his fellow men should perceive him acting thus.[1]

Here again is the notion that our insanguination as human beings is a pollution. We stain ourselves anew with animal sacrifice. The great difference between Homer and the Orphics turns on this: can one exculpate oneself from the crime of killing and eating an animal by sacrificing the animal to the gods? For the Homeric worshipper the answer to this question is yes; for the Orphic the answer is no. For the Orphic, indeed, the spilling of blood on the altar in the act of animal sacrifice is a hideous pollution. Far from purifying the killer, it adds sacrilege to the crime of murder. We felt the force of this when father sacrificed son with a prayer in the lines of Empedocles above. The prayer makes it especially bad. In his essay on abstinence from animal food Porphyry mentions the altar of the pious on the island of Delos, on which no animal was ever sacrificed.[2]

In the Homeric epics sacrificial killing is enjoined upon Odysseus by Circe who provides him with the creatures whose blood will attract the dead. Teiresias tells Odysseus of the animals which he must sacrifice to Poseidon when he finds the people who know nothing of the sea or ships. Alcinous and Aeolus, kings of two perfect kingdoms, rejoice in continuous feasting on meat. The gods themselves are accustomed to join Alcinous' court at the table though they do not eat the same food. But already in Hesiod's work the picture is a little different. Hesiod gives us a very uneasy story about the origins of animal sacrifice in which Prometheus tricks Zeus into accepting the inedible parts of the animal as his portion while the best of the animal goes to the sacrificer.[3] Zeus is angered by the trick and ill disposed to mankind in consequence, since it was on our behalf that Prometheus played the trick. This accounts, perhaps, for the practice of sacrificing and burning animals but only tasting them, so that the whole animal is truly the offering. This practice Porphyry ascribes to Pythagoras himself.[4] But there is one passage in

1. *Ancilla to the Presocratic Philosophers*, op. cit., Heraclitus, frag. 5.
2. Porphyry, *On Abstinence from Animal Food*, op. cit., p.82.
3. *Hesiod, The Homeric Hymns and Homerica*, op. cit., 'Theogony'.
4. Porphyry, *On Abstinence from Animal Food*, op. cit., p 82.

Homer where the sacrificing of animals for the purpose of eating them does turn to impiety: when the companions maintain their sad little ritual as they slaughter and eat the Cattle of the Sun. We feel keenly here the lack of aversive force in their actions, how their rags of piety make things worse. But this is not usually the case. When Odysseus and the companions enter the Cyclops' cave, they sacrifice one of the Cyclops' animals and eat it. This sharply contrasts with the Cyclops' own practice to Odysseus' credit.

But in the wider perspective, the perspective in which Polyphemus is the exact image of our own bestial humanity, every human being is there with the companions on Thrinacie. Like them we are damned if we eat and damned if we do not. As Eurylochus eloquently explains, we have the choice of death by starvation or of death by other means, and most other means are preferable to starvation. In the larger perspective we must choose between starvation and committing a crime. This crime binds us indefinitely to imprisonment in the cave as eater or eaten, to the disgusting belly.

How living the victims are in these stories! They wriggle and squirm but there is no pity. The companions who were eaten by Polyphemus are at least brain dead by the time they reach his mouth. But he keeps them fresh, he does not kill them all at the beginning. No such unconsciousness for the companions taken by Scylla who call out to Odysseus by name as they are whisked up to Scylla's cave, the most pitiful thing which Odysseus sees on his journey. Worse still the feast on Thrinacie where the Cattle of the Sun continued to low and crawl after they had been turned into beef and hides.

The predicament of the companions on Thrinacie and the entrapment of Odysseus in the Cyclops' cave are images of the doom which has overtaken every mortal creature. All are caught in the cycle of mutual killing and eating, being killed and eaten in ones turn. Living creatures are the redundant energies of other living creatures who have been killed or have died a little sooner. This is a bleaker vision of our animal existence than that of Darwin's struggle for life. Is there any relief from it? Odysseus escapes both eating and being eaten: on Thrinacie he simply abstains from even tasting the slaughtered cattle; in the Cyclops' cave he escapes by the expedients of calling himself No one; of blinding the Cyclops; of hiding himself and his companions under the sheep. If his imprisonment in the Cyclops' cave is symbolic of the animal and human condition, then these expedients of his also require a symbolic or anagogical interpretation which will relate them exactly to our own condition. We have seen our own predicament in the story of the Cyclops. The story may also tell us how to escape it.

* * *

When Odysseus told the Cyclops that his name was No one, the Cyclops never questioned its oddity. It must have struck him as a wonderfully appropriate name for the puny little runt in front of him. For Polyphemus size is the fundamental criterion for assessing the worth of man or god. Telemus may have been a good fortune teller, but he was also good and big, and when he prophesied to Polyphemus that Polyphemus would be blinded by Odysseus, Polyphemus supposed that this Odysseus would also be a fine, big fellow of greater strength than himself, not the feeble little nobody he actually turned out to be. Made drunk and blinded, in his extremity he calls out to the other Cyclopes. When they are standing outside his sealed cave, they ask him what is wrong and he tells them that No one has hurt him. They take him to mean that he has not been attacked and depart, leaving Polyphemus in a most marvellous confusion. How can the poor giant, befuddled, blind and maddened by pain explain to his fellow Cyclopes that when he said that No one had hurt him, he did not mean that he had not been assaulted by anyone, but that this man who was called No one...! If he was slow on the uptake before, he is completely immobilized in these circumstances. The logic required to analyze the complex wordplay defeats him.

Imagining Polyphemus' state of mind at this moment, the reader laughs along with Odysseus at the success of this improbable stratagem. Odysseus could not have been sure that Polyphemus would use the name No one when he replied to the Cyclopes' query. If Polyphemus had replied that pirates, say, had attacked him, then the Cyclopes would have come to his assistance. So Odysseus here is laughing partly in relief that his long shot had hit the mark. The first time reader, on the other hand, laughs in part because the point of Odysseus' calling himself No one has at last become clear. But both reader and Odysseus laugh here, too, at Polyphemus' discomfiture. We enjoy exactly that emotion which drove Odysseus to revile Polyphemus a few hours later and provoke the curse.

Odysseus laughs at his excellent stratagem. As D.C.H. Rieu indicates, the word which Homer uses for stratagem at this point is *mētis*, and this same word means 'no one'.[1] So Odysseus is also laughing at his clever pun on stratagem and 'no one'. It is all very amusing. But the last laugh here is certainly on Odysseus, in one of the most exquisite of Homer's ironies. If Odysseus had realized the value of calling himself No one, he would never have revealed his name to Polyphemus on his escape. He would have kept quiet and there is a good chance that Polyphemus would not have connected the feeble little nobody who had blinded him with the Odysseus whom Telemus had said would blind him. In that case whom would Polyphemus have cursed? No one? But Odysseus did not understand the depth of what he himself had said when he called himself No one in the Cyclops' cave. He thought it was just a clever trick. And so he told the Cyclops his real name on his escape, and in that name was cursed. Doomed to lose all his companions, his remorse destroyed the confident Odysseus who set sail from Troy. In Alcinous' court and in disguise in his palace in Ithaca he really becomes no one, through a terrible suffering which he need never have undergone if only he had realized the wisdom of his own words to Polyphemus all those years before. Like Achilles, Odysseus, too, could have left the battle and returned home safe if he had grasped the sense of what he had said himself. So Odysseus' clever trick, pun and all, is much funnier, in a very grim way, than even Odysseus appreciated at the time. But by the time he reaches Phaeacia and Ithaca he has learnt to preserve the secret of his identity to the point of obsession.

We must distinguish carefully between two different No ones.

---

1. Homer, *Odyssey*, tr. E.V. Rieu, revised by D.C.H. Rieu and Dr. P.V. Jones (Middlesex: Penguin Books, Ltd., 1991), footnote in revised edition, p136.

The No one who Odysseus ceases to be when he shouts his real name at Polyphemus is not quite the No one who escapes from Polyphemus' cave. For if that cave represents the necessity that restrains all mortal creatures to kill and eat, and to be killed and eaten, then the mere forsaking of our given name will not bring delivery from the cave. One must become No one more literally by transcending the condition of mortality altogether, by ceasing to be a creature at all. Odysseus becomes No one in both these different senses in the course of the *Odyssey*. He becomes No one in the first sense when he sheds the heroic values of honour and reputation and finds a refuge for his grief in passing among human kind without a name. He becomes No one in the second sense when he is offered immortality by Calypso and enters the southern gates of the cave of the Nymphs. He is, to be sure, still embodied; he has been liberated from the wheel of life and death while still living. So the first No one is anonymous, the second is immortal. To the first No one we may counterpoise 'being Odysseus'; to the second No one 'being a mortal'. These two No ones, though different, are related. The transcending of ones name and the need for other people's respect does not entail the transcending of ones condition as a mortal creature. But the transcendence of mortality would certainly entail the transcendence of the need for recognition in society. In the *Odyssey* these two kinds of transcendence are woven together from our first encounter with Odysseus until the end of the poem. We may call them the No one who is not Odysseus, and the No one who is not mortal.

At the time, then, when Odysseus tells Polyphemus that his name is No one, Odysseus has transcended neither his status as Odysseus nor his mortality. His calling himself No one is a clever trick which nicely gauges Polyphemus' contempt for his diminutive self and Polyphemus' incapacity for close logical analysis of a complex wordplay in an emergency. So Odysseus is only too ready to reveal his true name to Polyphemus on his departure. This revelation plays its part in the calling down of the curse upon them and leads directly to the companions' deaths. This is the burden of grief which Odysseus brings to Phaeacia and Ithaca, together with an obsessive desire to conceal his real name as much as he can. The last time we see this obsession at work is in his encounter with his ancient father Laertes. Through the course of the poem Odysseus learns the value of being the No one who is not Odysseus. His struggles with his own identity as Odysseus are a major theme of the poem.

The No one who is not a mortal is a much loftier No one than the No one who is not Odysseus. How far may we interpret Odysseus in the *Odyssey* as transcending his condition as a mortal creature alto-gether? Nothing less than this will secure his release from Polyphe-mus' cave, from the belly, from his own nature as a creature which kills and eats to survive. We know that he does merit immortality and we have explained that achievement as the reward for escaping the planetary whirlpool and for visiting the dead before his time. But can we see in Odysseus, as Homer describes him, an explicit transcendence of his condition as a mortal creature in the terms in which that condition is represented by Polyphemus and his cave? Here Odysseus' abstinence from the Cattle of the Sun will hardly serve because these are the only cattle from which Odysseus holds his hand and he has been warned by Teiresias and Circe to do so. But Odysseus makes some remarks to Alcinous which suggest a deeper level of self-hatred than the hatred of his identity as Odysseus. Alci-nous suggests to his court that the man before them, whom they do not know to be Odysseus, is one of the immortals, but one who is in disguise. Odysseus tells Alcinous not to entertain this notion, and he denies that he looks like the immortal gods who live in heaven.

> 'Alcinous,' Odysseus was quick to reply, 'on that score you may set your mind at rest. You can see that I have neither the looks nor the stature of the immortal gods who live in heaven, but am a human being. Think of the wretches who in your experience have borne the heaviest load of sorrow, and I will match my griefs with theirs. Indeed I think that I could tell an even longer tale of woe, if I gave you a full account of what I have been fated to endure. But all I ask of you now is your leave to eat my supper, in spite of my troubles. For nothing in the world is so incontinent as a man's accursed appetite. However afflicted he may be and sick at heart, it calls for attention so loudly that he is bound to obey it. Such is my case: my heart is sick with grief, yet my hunger insists that I shall eat and drink. It makes me forget all that I have suffered and forces me to take my fill. But at

daybreak I beg you to make arrangements for landing this unfortunate guest of yours in his own country.[1]

Odysseus does not directly deny Alcinous' suggestion that he is
an immortal, and a little later he tells Alcinous that he was offered
immortality by Calypso but turned her down. Odysseus may be
wrong when he says that he does not look like the gods. Nausicaa
has told her maids, out of Odysseus' hearing, that Odysseus does
look like the gods. And we know, as Nausicaa and the maids do not,
that Athene has transformed him after his bath. Of course Odysseus
may not still look like an immortal by the time he reaches Alcinous'
palace. Athene has made him invisible in the meanwhile and perhaps on his return to visibility in Alcinous' court he had shed his
divine appearance. But there must be some reason for Alcinous to
suggest that Odysseus is an immortal despite his anonymity, and
that must be his appearance, either its suddenness or the way Odysseus looks. Odysseus' denial is quick. He is quite unaware of the stir
which his appearance has caused.

Having disclaimed his immortality, if that is what he does, Odysseus calls himself a human being, the most miserable of them all.
Given Aeolus' rejection of him on his second visit, Odysseus is taking a risk when he reveals how unfortunate he is. Aeolus had
rejected him on his return to Aeolia on the ground that his misfortune with the bag of winds showed him to be a man detested by the
gods. Odysseus, having disclaimed that the Phaeacians owe him
respect as an immortal, is appealing now as the neediest suppliant.
But he has already presented himself in these terms to Nausicaa on
the shore and she has assured him that gods send trouble to mortals
whatever their merits. So he may be confident of the Phaeacians'
pity, especially as Queen Arete is among them. But then, with an
apology, he turns from them back to the food. If he has become
hateful to himself as the most unfortunate man in the world, he
becomes the more hateful to himself, he says, when his hunger
drives him to forget his grief as he restores himself. The belly, to use
his words, is incontinent and hateful because it distracts him from

---

1. *Odyssey,* VII, 208–224.

that grief. The more normal response to this capacity of the belly to distract us from grief is heartfelt thanks that for a moment the misery is lifted. Helen adds nepenthe to the wine to this very end. But here Odysseus condemns the belly for distracting him.

There is a disassociation here between Odysseus and one of his own organs in his criticism of the belly. It is as though he has been lumbered with it but it is not him. But to whom else does one refer when saying 'I am hungry' than ones self? Who is the Odysseus who can talk of his own belly in this way? The answer is clearly that the Odysseus who grieves is the real Odysseus while the Odysseus who hungers is shameless and hateful. We might compare Odysseus' disassociation from his own body to the words of Christ in the Sermon on the Mount:

If thine eye offend thee, pluck it out.[1]

The sense of a self apart from the body, in Odysseus' case the grieving self, may become so strong that it turns against the body and condemns or mutilates it. So we may say that the No one who is not a mortal does appear in the poem in the terms in which mortality is represented by Polyphemus in his cave. Odysseus does, on occasion, express himself as hating the belly to which he is tied as well as hating his identity as Odysseus among men. But he does not condemn the belly on Empedoclean grounds for the violence against kin and the bloodshed which it provokes. Odysseus condemns the belly because it makes him forget the grief which has become more himself than his own body. In Odysseus' case at least, his hatred of his identity as Odysseus has brought about transcendence of his identity as a mortal creature. His grief as Odysseus has become so deep and continuous that he hates his need for the food which will momentarily make him forget it.

Odysseus calls himself No one and learns later the full value of that anonymity. Later, too, he is offered immortality by Calypso and enters the southern gate of the Nymphs' cave. In many religious orders, the neophytes put away their own names and adopt new names, to signify the breaking with their own past lives. St Paul was

1. Matt. 5:29.

called Saul until his conversion on the road to Damascus. He is blinded and when his sight returns he is a new man with a new name who can say 'Not I but Christ in me.'[1] It is as though his previous character as Saul the persecutor has been removed to reveal the real being which that character had concealed. William Blake called his human self an incrustation over his immortal selfhood.[2] Similarly Empedocles' account of the great oracle describes the punishment meted out to those long-lived spirits who have stained their limbs with bloodshed or sworn a false oath.[3] These long-lived spirits are excluded from the company of the gods and assume all manner of earthly forms. But they are still spirits who will return to their proper state of being after thrice ten thousand seasons. In Hesiod and Homer, gods who swear oaths do so on the waters of Styx, and the price of forswearing such an oath is, as with Empedocles, exclusion from the company of the gods. For Hesiod the exclusion is for ten years only.[4] The first of these years is spent in a breathless stupor, the other nine in penance. The swearing of the false oath is common to Homer's and Hesiod's accounts and to the great oracle of Empedocles but we do not find in Homer and Hesiod the notion that this penance is exacted from the forsworn god in mortal sufferings. There is no indication in their accounts that human or animal life, the life of Odysseus for example, is the fallen condition of a god.

Calypso swears her oath to Odysseus on the waters of Styx, the most dreadful oath of the gods. But though Odysseus both calls himself No one and becomes immortal, there are few indications in the *Odyssey* that mortal creatures, even heroic humans, are fallen immortals. One of these indications may be discerned in the story of Circe who turns Odysseus' companions into pigs and shuts them in her sty, though they still have the minds of men. The condition of a man-pig here is comparable to a god incarcerated in a human form, a divine spirit 'clothed in the unfamiliar tunic of flesh' as Empedocles

---

1. Gal. 2:20.
2. *William Blake's Writings, Vol. I*, ed. G.E. Bentley Jr. (Oxford: Clarendon Press, 1978), 'Milton, Book the Second', plate 40, 35–36.
3. *Ancilla to the Presocratic Philosophers*, op. cit., Empedocles, frag. 115.
4. *Hesiod, The Homeric Hymns and Homerica*, op. cit., 'Theogony', pp 134–137.

puts it.[1] Another indication is the killing and eating of the Cattle of
the Sun on Thrinacie. These cattle are immortal and they go to
making up the bodies of the doomed companions who eat them. I
have suggested that doom in Book XII whether the Sirens, Scylla,
Charybdis or the thunderbolt is a return to the planetary whirlpool
which conditions all mortal existence. The immortality of the eaten
cattle is, as it were, the source of the life in the mortal creatures which
are precipitated from their slaughter. Since I have identified these
immortal cattle with the ever returning days of the year, it follows
that each of us is the fragmented energy of his or her natal day, that
one day of the year on which we celebrate our birthdays and which,
like every other day of the year, is an immortal.

But if Odysseus was not once an immortal who has fallen to
become Odysseus, how else are we to understand his becoming
immortal? If he had really become No one by the time he reached
Ogygia and is offered immortality, what else was there of him, when
his personal and creaturely identity had been discarded, which
could become immortal? It is highly significant that his entry
through the southern gate of the cave of the Nymphs was accompanied by Athene. We may assume that he does indeed enter through
the southern gates, a detail which is not entirely clear in the poem,
just because Athene is with him and the southern gates are the gates
of immortals like her. Athene had always looked after Odysseus at
Troy but when he fell foul of Poseidon, she would do nothing to
offend her uncle. Poseidon's absence from Olympus enabled her to
intervene on Odysseus' behalf after his many years on Ogygia. She
feels that their two personalities are entirely consonant. She is the
arch schemer in heaven and he on earth, she tells him. If there is a
divine or immortal dimension to Odysseus which transcends his
personal and creaturely identities, this dimension is his cunning.
But it is not *his* cunning any more for he has become No one. This is
the man-goddess who has returned to Ithaca.

* * *

1. *Ancilla to the Presocratic Philosophers*, op. cit., Empedocles, frag. 126.

The other horrific passage in the Cyclops episode, after the eating of the companions, is the blinding of the Cyclops:

> I went at once and thrust our pole deep under the ashes of the fire to make it hot, and meanwhile I gave a word of encouragement to all my men, to make sure that no one should play the coward and leave me in the lurch. When the fierce glow from the olive stake warned me that it was about to catch alight in the flames, green as it was, I withdrew it from the fire and brought it over to the spot where my men were standing ready. Heaven now inspired them with a reckless courage. Seizing the olive pole, they drove its sharpened end into the Cyclops' eye, while I used my weight from above to twist it home, like a man boring a ship's timber with a drill which his mates below twirl with a strap they hold at either end, so that it spins continuously. In much the same way we handled our pole with its red-hot point and twisted it in his eye till the blood boiled up round the burning wood. The fiery smoke from the blazing eyeball singed his lids and brow all round, and the very roots of his eye crackled in the heat. I was reminded of the loud hiss that comes from a great axe or adze when the smith plunges it into cold water - to temper it and give strength to the iron. That is how the Cyclops' eye hissed round the olive stake. He gave a dreadful shriek, which echoed round the rocky walls, and we backed away from him in terror, while he pulled the stake from his eye, streaming with blood. Then he hurled it away from him with frenzied hands and raised a great shout to the other Cyclopes who lived in neighboring caves along the windy heights.[1]

In the Aeolus episode the bag of winds and its silver string represent the human breast and voice. In the Cyclops episode Polyphemus' cave represents the human belly. In the account of Polyphemus' blinding we are given, I think, a symbol of sight and of the human eye. Eye, indeed, rather than eyes, for though Polyphemus has brows he has but the one eye. The lance in this operation is cut from Polyphemus' staff of olive wood, not yet seasoned, which to Odysseus and his companions looks like the mast of a ship with twenty oars. It is most unusual to find a piece of olive wood which is straight enough to make a staff even for a normal human. But perhaps green olive wood is very hard to burn and is capable of being

---

1. *Odyssey,* IX, 375–400.

heated to the point which Odysseus describes. Why is the stake heated? It must have made it harder for the companions to handle and it is not clear that it would have had greater penetration because it was hot. But the red-hot stake seems to be a feature common to the many forms which this tale takes in the folkways of nations. Sometimes the stake is the red-hot spit on which the giant has been roasting his victims.

But if the blinding here is a symbol of physical sight, then the heating of the stake until it glows is quite comprehensible. The stake represents a beam of light, that peculiar additional requirement for any act of seeing to which there is no equivalent in the operations of the other senses. But seeing needs light, and the light here is represented by the red-hot stake. So sight here, the meeting of eye and light, is symbolized paradoxically by the process of being violently blinded. Consider for comparison the way in which the Buddhist Thankas of Tibet represent sight among the five senses. It is represented on an inner ring of the great wheel of life held between the jaws of Yama, the god of death. Sight is a man, often kneeling, with head upraised and an arrow fixed in his eye. Here is the stake but not the glow, though the arrow may represent the swiftness of light instead. Or consider these lines of Andrew Marvell in which the soul complains to the body about the pains to which the body subjects it:

> Here blinded with an eye; and there
> Deaf with the drumming of an ear,
> A soul hung up, as 'twere, in chains
> Of nerves, and arteries, and veins. . . .[1]

From this point of view our faculty of physical sight actually blinds and blocks the soul just as hearing deafens it. The soul is invisible and inaudible in its own nature and is at its best when wholly ingathered upon itself. The physical senses are like wounds in its integrity. Or the ingathered soul may be represented as the soul awake and fully self-conscious, while the soul which perceives the physical world is in a trance or asleep.

---

1. Andrew Marvell, *The Complete English Poems*, ed. Elizabeth Story Donno (London: Allen Lane, 1974), 'A Dialogue Between Soul and Body', p103.

But this way of thinking is again Platonic rather than Homeric. There are no souls in Homer's epics like the immortal souls of whom we read in the works of Plato and Empedocles. The nearest explicit approach to such a notion in the Homeric epics is the account of Heracles in the book of the dead. After death, Heracles himself is with Hebe among the immortals while his wraith is in Hades. It was, no doubt, because this notion of an immortal Heracles on Olympus does not fit with Homer's other accounts of the dead in the *Iliad* and the *Odyssey* that the lines describing Heracles in heaven after death were omitted by some ancient commentators and textual critics. The notion that our physical senses blind us to immaterial, non-physical reality is one of the guiding principles of Platonism, but we are given no indications of an immaterial realm in Homer, except perhaps that of the gods. And even if the gods are finally immaterial, there is no notion in Homer that we have a special organ for comprehending these immaterial entities such as Plato supposed *nous* or intellect to be. And lastly it is not Odysseus who is blinded here but Polyphemus. He would be the one to benefit from his sudden projection into the invisible, if such existed for Homer, not Odysseus.

But in the same way that Polyphemus' cave is Odysseus' belly, Polyphemus' eye is Odysseus' eye. The cave, as it were, is Odysseus' belly turned inside out so that it now holds Odysseus as its victim. That, I have argued, is the point of making Polyphemus a human being, so that we may experience what our victims suffer at our hands as though we were they. Polyphemus, in fact, is a novel way of presenting our own bestial humanity to us. The cave is his belly but it is also ours. In the same way his eye is our eye, a human eye, and I defy anyone to read the description of Polyphemus' blinding and not feel it as though suffering Polyphemus' pain. Here the enormous size of the giant's eye corresponds exactly to the experience of intense pain in such an organ, pain which seems to expand so as to occupy the entire consciousness. And we are given not just a blinding but an anatomy lesson. The poet explains exactly what happens to the optic nerve in these circumstances. This involuntary imagining of Polyphemus' pain on the part of the reader marches oddly with the desire to see Odysseus and his companions escape.

Metaphorically Odysseus becomes No one in order to escape the human state represented by the cave. He must blind Polyphemus to the same end. His escape is from both the belly and the eye, the two organs of fire in the human body. His calling himself No one and his blinding of Polyphemus are not merely coincident, but parallel or congruent. The nonentity of not having a belly and the nonentity of not seeing point in the same direction, towards a state of being which we may call trans-human. This is the state which Odysseus symbolically gains and then loses by taking back his name.

Blindness is an impairment of both Demodocus and Teiresias in the *Odyssey,* and Homer himself was generally believed to have been blind. It is very hard not to connect the extraordinary capacities of these people with their blindness. Their visionary powers, we imagine, must have been freed by their loss of physical sight. But blindness is neither a necessary nor a sufficient condition of prophetic or poetic vision. It is not necessary because many poets and prophets are sighted, and it is not sufficient because many blind people lack visionary power. And yet our minds still make the connection, perhaps because the very act of reading or hearing the *Odyssey* conjures that invisible world for our inner eye while our physical seeing has almost ceased to engage our awareness. But the blinding of Odysseus, if that is what it is, is much more like the self-blinding of Oedipus, than the occlusion of physical seeing occasioned by vivid imagining. Odysseus comes to loathe himself as Odysseus and to loathe the self which is bound into the murderous cycle of the food chain. If the blinding is of a piece with this, then he comes to loathe this bond, too, to the animal self. Plato called the physical senses the nail which fastens soul to body. Oedipus blinds himself at the end of Sophocles' play because he can no longer bear to look upon his children by the same mother as himself. He suffers a terrible revulsion from the visible, and cannot bear to look at it. On this reading, the cave of Polyphemus from which Odysseus escapes is a symbol of the whole physical realm, like Plato's cave in the *Republic.* Plato's prisoners need not blind themselves in order to escape the cave, but they must turn away from the spectacle on its back wall, and be thought no better than blind for having done so by those who have not.

The Cyclops encounter is the first of Odysseus' adventures to which a full-length narrative is given. Compared to the length of this narrative, the accounts of the Cicones and the Lotus-eaters are cursory. The Cyclops encounter is therefore the grand overture to the sequence of the adventures. Odysseus is trapped in Polyphemus' cave, escapes and is trapped again by an even worse doom than being eaten by Polyphemus. For his second entrapment destroys all his companions and not merely the ones with him in the cave. Few of us suffer the misfortune of killing hundreds of our followers through our own stupidity. To this extent the story of Odysseus' fall and redemption is peculiar to the single individual Odysseus. But the long story which opens the sequence of Odysseus' peculiar destiny, as that destiny is related in the *Odyssey*, is the common adventure of all creatures and certainly of all humans. Every human by being human here on earth is bound into the struggle for life, and Polyphemus represents that aspect of ourselves. We may escape from this, as Odysseus does, by becoming No ones and by blinding the eye of physical sight. So this episode within the story of Odysseus' peculiar path to salvation applies equally well as it stands to everyone without exceptions. It certainly applies to Odysseus at this other level, because he is embellied like us and knows it and talks about it in these terms. The Cyclops episode provides the root or base of our condition as creatures and humans. It is the filthy, violent starting point from which the journey to redemption begins.

6

THE STY OF THE WORLD

CIRCE'S DRUGS are uniformly effective in changing the men who ingest them into wolves or lions or pigs until Odysseus successfully resists them. And Circe may have been efficient in her other purpose with these drugs: to make those who take them utterly forget their native land. But it appears that those who take them retain the minds of men even when incarcerated in the bodies of animals. We are told this of the companions who are turned into pigs and we infer it of the wolves and lions who guard Circe's palace, from their extraordinary playing and gambolling with the companions. There are men in those animals.

Is this possible? Could the mind of a man find itself in the body of an animal? This is how Homer describes the metamorphosis:

> Circe came out, opened the polished doors, and invited them to enter. In their innocence, the whole party except Eurylochus followed her in. But he suspected a trap and stayed outside. Circe ushered the rest into her hall, gave them settles and chairs to sit on, and then prepared them a mixture of cheese, barley-meal, and yellow honey flavoured with Pramnian wine. But into this dish she introduced a powerful drug, to make them lose all memory of their native land. And when they had emptied the bowls in which she had served them, she struck them with her wand, drove them off, and penned them in the pigsties. For now to all appearance they were swine: they had pigs' heads and bristles, and they grunted like pigs: but their minds were as human as they had been before the change. Indeed, they shed tears in their sties. But Circe flung them some mast, acorns, and cornel-berries, and left them to eat this pigs' fodder and wallow in the mud.[1]

1. *Odyssey,* x, 230–243.

In the case of the companions we might suppose that it would only be a matter of time before their new bodies as pigs and their living conditions obliterated their memories of themselves as human beings and they became entirely animal. But this clearly has not happened with the wolves and lions. Or perhaps this is just where Circe's drugs are most fiendishly effective, leaving their victims with an undiminished sense of their own degradation. On the other hand we may say this for Circe, that where her drugs to turn men into animals do not obliterate their human mentality, her drugs to turn those animals back into humans more than restore them to their former condition. As a result of their two metamorphoses Odysseus' companions looked younger and much handsomer and taller than before. The process had provoked their anxiety but the results must have made it almost worthwhile.

Plato has views on the relations between the human mind and the human body. He supposes that the human head is the supreme achievement of the sublunar creation because it most closely resembles the heavens. The cranium is a hemispherical dome and the single band around the head which covers the eyes and ears corresponds to the tropic zone within which the sun, moon and planets run their courses. The human senses of sight and hearing are perfectly adjusted to seeing the courses of the heavens and hearing the musical intervals which organize the planets and consciousness itself. The human body is merely the stalk or vehicle which supports this wonderful head, and it holds the head at its summit, the better to see the heavens and to govern the whole creature. Since the human head is the most perfect creaturely creation, it follows that all other animal forms are devolutions from it and must be understood as more or less distant from this human paradigm:

> Land animals came from men who had no use for philosophy and never consider the nature of the heavens because they have ceased to use the circles in the head and follow the leadership of the parts of the soul in the breast. Because of these practices their fore-limbs and heads were drawn by natural affinity to the earth, and their fore-limbs supported on it, while their skulls were elongated into various shapes as a result of the crushing of their circles through lack of use. And the reason why some have four feet and others many was that

the stupider they were the more supports god gave them, to tie them more closely to earth.[1]

If the operations of the human mind imitate the revolutions of the heavens and if the roundness of our human heads enables these revolutions in us to a greater degree than is possible in quadrupeds, then it is hard to see how a human mind could last for long in the elongated head of a quadruped. But of course Plato is not talking of a Circean metamorphosis here, of an actual man into an actual quadruped. He is speaking morphologically. He is describing analogies between the forms of various species of animals, and he locates the origin of these variations in their capacities for philosophy and astronomy. The physical frame is made for the task which the mind has set itself. What then of Odysseus' companions who are turned into pigs, and of Odysseus himself who is not? Plato's devolutionary morphology has no place here, but we do wonder whether the companions somehow deserved their being turned into pigs. Yet we are told that Circe's magic has never failed of its victims before Odysseus. Every man she had encountered she had bewitched. Then again, Odysseus dryly remarks on the innocence of the companions who walked into Circe's trap, an innocence not shared by Eurylochus who waited outside. What of Odysseus? Was he exceptional in some way, which saved him from Circe's spell? Plato, it seems thought that he was. In his reworking of the Circe episode in his *Myth of Er*, Plato represents Odysseus as exceptionally wise.

We have already examined the astronomy of the *Myth of Er* in the second chapter, where we took Plato's account of the spindle on the knees of Necessity as the model or pattern which helped to explain Odysseus' adventures between Aeaea and Ogygia. To this we are helped by Plato's naming of the Sirens in his myth and by the explicit contrast he makes between his story and the story told by Odysseus in Alcinous' court. But the *Myth of Er* is concerned less with astronomy than it is with the destinies of souls, and their experiences as they pass into another mortal life. Towards the end of the myth, the souls who are about to be born again choose the next lives

1. Plato, *Timaeus and Critias*, tr. H.D.P. Lee (Middlesex: Penguin Books, Ltd., 1965), 'Timaeus', 44Dff, 91E.

which they will lead, and there is a general metamorphosis as the souls who were once people become other animals and those who were animals become people. This passage is certainly comparable to Homer's account of Circe, even though there is no explicit mention of Homer's goddess in the myth, at least under the name Circe. And of course there is no explicit account or suggestion in Homer's poem that the souls of the dead whom we meet with Odysseus will ever be reincarnated to live another life on earth. Homer's dead may be as anxious to drink blood and join the living as ever Circe's lions and wolves and pigs are anxious to rejoin the human race. But there is nothing in Homer's epics to suggest the cycle of life and death which we find in the *Myth of Er*. Homer's ghosts have the same name, *psychai*, as Plato's souls but nowhere does Homer suggest that they are reborn.

And yet, despite this difference, we are in Homer's world again as we see Plato's souls choose their next lives. Here again are Agamemnon and Ajax whom we saw as ghosts among the dead in the *Odyssey*. Here are Epeius and Thersites, who also served with Odysseus at Troy, and Odysseus himself, who is last to choose and whose choice is the climax of the myth. Do we have here Plato's way of understanding what Homer himself meant, or do we have merely Plato's creative reconstruction of Homer? Whichever is true, the *Myth of Er* provides us with an extended examination of the Circe episode in Homer.

> Most curious, he said, was the spectacle—sad and laughable and strange; for the choice of the souls was in most cases based on their experience of a previous life. There he saw the soul which had once been Orpheus choosing the life of a swan out of enmity to the race of women, hating to be born of a woman because they had been his murderers; he beheld also the soul of Thamyras choosing the life of a nightingale; birds, on the other hand, like the swan and other musicians, wanting to be men. The soul which obtained the twentieth lot chose the life of a lion, and this was the soul of Ajax the son of Telamon, who would not be a man, remembering the injustice which was done him in the judgement about the arms. The next was Agamemnon, who took the life of an eagle, because, like Ajax, he hated human nature by reason of his sufferings. About the middle came the lot of Atalanta; she, seeing the great fame of an athlete, was unable to

resist the temptation: and after her there followed the soul of Epeus
the son of Panopeus passing into the nature of a woman cunning in
the arts; and far away among the last who chose, the soul of the jester
Thersites was putting on the form of a monkey. There came also the
soul of Odysseus having yet to make a choice, and his lot happened
to be the last of them all. Now the recollection of former toils had
disenchanted him of ambition, and he went about for a considerable
time in search of the life of a private man who had no cares; he had
some difficulty in finding this, which was lying about and had been
neglected by everybody else; and when he saw it, he said that he
would have done this had his lot been first instead of last, and that he
was delighted to have it. And not only did men pass into animals, but
I must also mention that there were animals tame and wild who
changed into one another and into corresponding human natures—
the good into the gentle and the evil into the savage, in all sorts of
combinations.[1]

In the Circe episode we see men who have been changed into
wolves, lions, pigs and perhaps a stag. Circe tethers a young ram
and a black ewe by the ship as Odysseus is about to leave on his visit
to the dead. These two animals were to provide the sacrificial offerings and the blood for the invocations of the dead. Were they a
young man and a woman before meeting Circe? In Plato's account
we have a lion, swans, an eagle, a nightingale, and a monkey; we
have a woman, Atalanta, who becomes a man; a man, Epeius, who
becomes a woman; and Odysseus who is the only one to remain the
same gender and species. Plato's range is wider than that of Homer
who confines himself to quadrupeds as creatures other than
human. But Plato's range is narrower than that of Empedocles who
declares:

> For already was I born a boy and a girl, a plant, a bird and a dumb
> seafish.[2]

In the Empedoclean account of transmigration, the bodies
through which the long-lived spirits must pass during their ten
thousand year exile from the gods seem to be of all living kinds. In

1. Plato, *Republic*, op. cit.
2. *Ancilla to the Presocratic Philosophers*, op. cit., Empedocles, frag.117.

his Oracle of Necessity the spirits are described as successively inhabiting and being rejected from the four elements one after the other, as though their long journeys must each encompass all the possibilities of creaturely existence. This may, for all we know, be true of Plato's souls too. Plato tells us that the souls of singing birds may take on human bodies in their next lives, but it is hard to see how this would fit with his spiritual morphology of the animal heads in the Timaeus.[1]

For the most part the *Myth of Er* respects the morphology. The emphasis is always on the spiritual determination of the physical form, the way in which virtue determines ones next life form and destiny. This emphasis is very like Leibniz's emphasis on the priority of the kingdom of minds over the kingdom of bodies in his *Principles* and *Monadology*. The physical realm is exactly attuned to the moral destinies of souls, their salvation and return to god; the laws of physics and every individual physical event conspire to achieve this deeper purpose. In the same way Plato repeats several times that the choices of lives are made by the souls themselves, without compulsion, and that there are many more lives to choose from than there are souls to choose them. This certainly does not sound like the experience of the companions in Circe's palace. Like all men before them there, they are bewitched by an overwhelming potion and a spell, and there is little reason to think that they chose, deserved or incurred their fate. Indeed, there is more reason to call Circe Necessity than there is to call by that name the Goddess on whose knees the spindle of the universe turns in Plato's story.

How can Er tell who each of the souls is as they go forward to choose their next lives? Their names are never called, and since the group of souls contains the souls of animals who will choose to become human, personal names might not always be available. Clearly these souls are in some sense perceptible to Er, and in some way he is able to identify them but we are not told how he does this. When Odysseus sees the dead in book XI of the *Odyssey*, he either recognizes them because he knew them in the flesh, or they bear some sign like the golden rod of Teiresias, or they announce their

---

1. Plato, *Timaeus,* op. cit., 91–92, and *Republic,* 620.

lineage as the women do who are sent up by Persephone. Perhaps Er came to know the identities of some of his band during the seven days on the meadow or on the four-day journey which followed. For all that, these Platonic souls in the *Myth of Er* are rather different from the souls discussed elsewhere in the *Republic,* with their three parts or in their ascent from the cave. However Er perceives these souls, they are still highly individualized and seem oddly Homeric. They are very like the wraiths whom Homer calls souls in the *Odyssey,* but the wraiths represent the humans who they once were at the moment of their deaths, 'with their spear wounds gaping yet.' Plato's souls in the *Myth of Er* are not so described. Their thousand years in heaven or below the earth seem to have restored them. It is as though Plato has temporarily adopted the Homeric soul for his story, and then adapted that soul to his reincarnationist vision. Plato's souls are much more active than Homer's but they are not by any means the immanent principles of life which they are in the *Phaedo.* As a result his account of the soul is not entirely consistent.

The description of the soul's choosing in the *Myth of Er* changes as it proceeds. At first the souls of Orpheus, Agamemnon, Ajax choose their lives from those laid on the ground before them, but as the list proceeds, the souls actually change into their next embodiments before us. Epeius, the bone crushing wrestler of the *Iliad* and architect of the wooden horse, passes into the nature of a woman cunning in the arts. Thersites' soul we actually see putting on a monkey. Homer had described this Thersites in the *Iliad* in the following terms:

> He was the ugliest man to have come to Ilium. He had a game foot and was bandy-legged. His rounded shoulders almost met across his chest; and above them rose an egg-shaped head, which sprouted a few short hairs.[1]

As Thersites puts on his monkey we are imagining his more or less Homeric wraith actually in the process of transformation where we can mark the physical resemblance between what he was and what he has become. Perhaps we can see the old Thersites peeping

---

1. *Iliad,* II, 216–219.

out of the monkey's eyes, as Homer mentions the companions' tears in the eyes of pigs.

Thersites is late in the queue but Odysseus is last of all to choose, for so the lot had fallen. Plato describes Odysseus at this point in his career at some length, but his is the only soul to choose a life of the same species and gender as his last one, so there is no fascinating metamorphosis for him. His persistence in the form of a man, according to Plato, repeats his achievement in Circe's palace where he was the only man of his crew not to succumb to Circe's transformative magic. Circe tells him that he is the only man ever to have resisted her spell. So Plato's account of Odysseus' choice corresponds to Homer's presentation of him in Circe's palace. According to Plato, by the time Odysseus comes to choose his next life, the recollection of his former toils has disenchanted him of all ambition. If this is how Plato supposes Odysseus to be after the rest of his life as Odysseus and after a thousand years in the other world, then we would have to say that Odysseus has changed very little from the man Homer described as returning to Ithaca. The grief is still there, untempered, the coal at the core of his being. And he is still disenchanted with ambition and the love of honour. This is the same man whom we see weeping in Alcinous' palace as Demodocus sang of his exploits at Troy. So far Plato has not developed the character of Odysseus beyond the point at which Homer leaves it. His description accurately depicts the Odysseus of the *Odyssey* in Phaeacia and Ithaca.

Odysseus, according to Plato, goes around for a considerable time looking for his next life. He seems to be under no pressure and to be much more careful than the souls who preceded him. His being last in the queue does not seem to have disheartened him. He has heeded the prophet's words that even the last comer will be happy in the next life if the choice is wise. This, too, is exactly the Odysseus of the *Odyssey* after the sharp lessons of the Cicones, the Cyclopes and the Laestrygonians. We might imagine those lives on the ground in front of Odysseus as the opportunities which he passes up in the *Odyssey*: a life with Circe or with Calypso or with Nausicaa as king of Phaeacia. The long time which Odysseus spends in making his choice repeats the nine-year journey which Homer describes. Of course we may also ask why Plato has Odysseus choose another life at all, if Homer's Odysseus has achieved immortality in the *Odyssey*. We might just accept his returning to Ithaca despite the offer of immortality and life with Calypso, but we cannot explain why in this case he would have to choose a whole new life on earth, after his death as Odysseus. But like the Odysseus who returns to Ithaca when he need not, and who risks Circe's spell when he need not, Plato's Odysseus makes the most selfless choice of those available to him.

At length Odysseus finds the life he has been looking for, lying spurned by all the souls that have chosen before him. When he finds it, he is delighted and declares that he would have chosen the very same life if he had been the first to choose. It is the life of a private man, removed from the cares of public life. Here at last, we may feel, is an Odysseus who is not Homer's Odysseus. For Homer's Odysseus returns to Ithaca fully intending to re-establish his rightful sovereignty at the very center of his kingdom's affairs. But just as Plato may have discerned a disenchantment with ambition in Homer's Odysseus, here too he may be ascribing to his Odysseus what he has seen in Homer's. And I think that he is right if he supposes that the Odysseus who returns to Ithaca has no real desire to govern for his own part, but sees it as a burdensome duty. This is Odysseus the champion ploughman, who enjoys pig farming with Eumaeus and discussing his orchards with Laertes. Certainly Plato's description of Odysseus' choice is far closer to Homer's Odysseus than are the poems of Dante,[1] Tennyson,[2] Cavafy,[3] and A.D. Hope.[4] These poets represent the returned Odysseus as dissatisfied with "the petty kingdom he called home." In their view Odysseus would have found it difficult or impossible to narrow his focus again after such excitement and experiences. Homer's Odysseus is a much more tormented man than theirs and he would be free of his responsibilities to all but a very few. This is the best of kings, a man

---

1. Dante, *Inferno*, Canto 26.
2. Tennyson, *Ulysses*.
3. Cavafy, *Ithaca*.
4. A.D. Hope, *The End of a Journey*.

who in his heart of hearts wants something simpler than to rule and who takes his responsibilities very seriously.

I have said that Odysseus' choice is the climax of the *Myth of Er* and that Plato supposes Odysseus to be the wisest of men. In the telling of the myth Socrates introduces a new definition of philosophy. Philosophy is what will enable one to make the best choice when ones turn comes to choose the next life. Though Odysseus is last in the queue he takes the greatest care in choosing, and when he has chosen what the others have left lying there, he says that he would have chosen the same if he had been first. This implies that Odysseus knew what he was looking for before he started. He was not concerned with what was on offer with this single exception. And there it was. Furthermore the life he chooses is described in terms that fit Socrates' own life. Socrates too kept out of politics and the law courts and all the forms of democratic or oligarchic government as far as he could. Odysseus' carefulness, his preparedness and his choice of life are characteristics of the *Odyssey's* Odysseus and they must all commend themselves to Socrates as the very type of the philosophical personality. But this is Homer's Odysseus who is standing before us as the last paradigm of philosophy before the *Republic's* end. Are we to imagine that Socrates is Odysseus' reincarnation, that this is the reason for Odysseus' rebirth? Pythagoras is reported to have had complete recall of his past lives and believed that one of his incarnations was mentioned in the *Iliad*.[1]

But even if Plato's Odysseus is Homer's Odysseus, it is hard to accommodate the vast machinery of Plato's other world in the *Myth of Er* to the little island of Homer's Circe. Aeaea is so small that Odysseus can see the palace at its center from a rugged height quite near the shore. Plato's other world is so vast that the Goddess Necessity holds the visible universe like a spindle on her knees. Circe deals with a score of companions whom she drugs and strikes with her wand, one by one. Necessity presides over the births and deaths of the whole world, over all the metamorphoses of all the souls in the cosmos. In the last chapter I generalized or universalized the cave of Polyphemus so as to represent the essential embelliedness of all

1. Porphyry, '*The Life of Pythagoras*', op. cit., 26, 45.

animal life. Did Plato believe that Homer intended the Circe episode to bear a similar meaning? Or is his remake of the Circe episode in the *Myth of Er* a creative extrapolation of his own?

Homer's Circe is much more than a witch who transforms Odysseus' men into pigs and back into men. She is the single greatest helper whom Odysseus meets on his travels. She knows exactly what his destiny requires of him and she helps him to achieve it. She knows that he must visit Teiresias among the dead and she knows exactly what Odysseus must do to consult him. Her directions to the region beyond Ocean stream are detailed and accurate. When Odysseus and his companions return from the dead to Aeaea in perfect safety, she is the one who informs Odysseus of the next stages in his journey, again in great detail and with perfect accuracy. Homer's Circe has three separate powers in the *Odyssey*: the power of metamorphosis over other beings; access to the realm of the dead; knowledge of the passage over the heavenly circuits represented by the Sirens, Scylla, Charybdis and Thrinacie. In this way the second half of book X and the whole of books XI and XII unfold under the aegis of this single goddess. She is the single most commanding presence in Odysseus' journey.

These same three powers or realms are under the authority of Plato's Goddess Necessity. Here again are the wraiths of the dead whom we met in the *Odyssey*. Here too are the metamorphoses of Circe's palace, and the whirlpool of the heavenly circuits, now spinning before the souls in a most glorious vision. Plato's *Myth of Er* recreates these two and a half books of the *Odyssey* synchronically. Instead of reading about Circe's transformings in book X and the heavenly movements in book XII, the spells and movements are coordinated in a single narrative. It is as though the Circe episode and the adventures of books XI and XII have been dissolved into each other, homogenized. On this larger view Plato's synthesis is much closer to his Homeric original, and the *Myth of Er* becomes a profound meditation on the larger part of Odysseus' narrative. Plato recognized depths in Homer which we do not. Socrates' disavowal at the beginning of the *Myth of Er* that his story is one of those which Odysseus told to Alcinous may spring from modesty, a sense of his own prosaic shortcomings.

Who are those people who descend through the northern gates of the cave of the Nymphs to be incorporated on the looms of stone? Homer does not tell us where they come from, but Plato does. According to Plato they are the recycled souls of the dead who have spent a thousand years outside the world. Is this what Homer supposed, or has Plato co-opted the Homeric stories to support an Orphic vision quite different from the poet's own? In the horror which Homer excites by his account of the Cyclops, we have an account of eating comparable to those of the Orphic Empedocles in his *Purifications*. In the urgent desire of Homer's dead to drink the blood and resume a surrogate life, we have the motive energy towards rebirth which we perceive in Plato's ascending souls as they anxiously cross the bellowing mouth which might yet deny them life. In Homer's account of the companions' eyes as they weep in Circe's sties we are given an uncanny sense of the interchangeability of human and animal bodies, of someone's looking out of something else's eyes. Pythagoras once told a man to stop beating a dog because in its bark Pythagoras could hear the voice of a friend.[1] Perhaps Odysseus could still distinguish among Circe's pigs the lineaments of his friend Polites.

* * *

There is another way of reading the Circe episode which saves us from taking Plato's path in the *Myth of Er*. This reading, too, turns on the companions' metamorphoses into swine. For the companions are not only changed, they are debased by becoming pigs. They are less than they were for all that they still have the minds of men. Their incorporation into pigs is a descent into a lower form of life, from which their mental life elevates them a little. It is an incarceration in a lower body. This parallels the Cyclops episode in which Odysseus escapes from embelliedness by becoming No one and then reassumes a creaturely life with his name when he reviles Polyphemus. We have another image of incorporation in the cave of the

---

1. Diogenes Laertius, *Lives of Eminent Philosophers*, tr. R.D. Hicks (London: Loeb Classical Library, 1925), 'The Life of Pythagoras', xx.

Nymphs where the creature is embodied in the marvellous fabric of sea purple on the looms of stone. But it is very difficult to represent the emergence of the formless into physical manifestation since the starting point is beyond representation. In the Cyclops episode formlessness is signified by the name No one; in the Ithacan return by the immortalized Odysseus and the companionship of Athene; in the Circe episode the fall into creaturely life is represented by the debasement of men into pigs. Here the human state represents the unfallen in comparison to the porcine. Pigs are to men as men are to what they were before becoming creatures. Homer's Circe is not quite the power who spins nature's wheel; she is the power who brings a human existence into its natural being from the beyond.

In the Polyphemus episode we feel a profound disgust at our own animal natures, at the horrific violence against other living creatures which sustains us. Our own bellies are represented by the filthy cave. In the Circe episode we are made to think of our own human bodies as if they were the bodies of pigs, which of course they are very like. This is the reason why Homer chose to turn the companions into pigs rather than into wolves or lions, not because of some characteristic particular to the companions, but because a furless pig with a rounded head is very like a human. Homer closely details what happens to their skin in these metamorphoses. The Polyphemus and Circe episodes both provide us with unfamiliar perspectives on our physical selves, which are all too familiar to us. But the Circe perspective is even cleverer than those of the cave and the giant. Homer imagines for us the process of our being turned into pigs and our actually being pigs as a way of making us feel what it is like to be in our own bodies as they are. But the bewitched companions still have the minds of men and yearn to be restored to their pristine state, just as we, in bodies like theirs, aspire to a reality finer than the brutal world into which we have fallen. Homer's story of Circe's spell is a clever device to make us feel afresh the magical, inexplicable fact of our own human embodiment.

The governing notion in Homer's account of Circe is not the metamorphoses of human to animal or animal to human. It is the notion of incarnation itself. The metamorphoses from man to animal and back again are symbolic and analogical, not literal. The turning of Odysseus' companions into swine represents analogically the turning of the superhuman into the human, the uncreated into the creature. The Circe episode is not concerned primarily with metamorphoses between states of creaturely existence but with the formation of a human being from a supermundane state. When Glaucon in the *Republic* hears Socrates describe the underground cave where prisoners are bound to posts from birth, he exclaims that it is an absurd image and they are absurd prisoners.[1] They are like us, Socrates drily replies. The spelling of Odysseus' companions, their turning into pigs and their weeping in the sties, this too is an absurd image and they are absurd prisoners. And again they are like us. We are the pigs in Circe's sties.

Odysseus is on the shore with the remaining companions when Eurylochus comes back to the ship with his dreadful news. His half of the one remaining crew has disappeared inside Circe's palace. They had all been mourning the loss of over nine tenths of their comrades, and now of the pitiful remnant left, a further half was lost. But Odysseus had had no choice. He had to know where Aeaea was if he was to steer a course home, and since the sky signs were unrecognizable, he had to find someone to ask. He had taken all reasonable precautions but he had failed again. Eurylochus refuses to guide him back to the palace, but argues for leaving the island at once with the crew who are left to them. He weeps and clasps Odysseus' hands as he begs not to return to the palace. Odysseus tells him to stay where he is but says that he himself must go, for a strong necessity is laid upon him. And so he goes. For Odysseus it is, I think, close to a suicide mission. He has lost so many and failed so badly that he would rather be dead than endure the loss of these men also.

As Odysseus makes his way through the sacred groves and enchanted glades of Aeaea, he is met by the god Hermes, looking like a young man, who takes Odysseus by the hand and asks him where he is going. Before Odysseus can answer, Hermes tells Odysseus that his lost companions are penned as pigs in Circe's sties. This is the first that Odysseus has heard of his comrades' fate. But

1. Plato, *Republic*, op. cit., 515.

Hermes continues, asking Odysseus if he intends to free them and warning that he too will fail to return and will remain with the others. Here, I think, Hermes does pause, playing with Odysseus to gauge the effect of his words, which are stunning enough. Having had his fun and having tested Odysseus' mettle, he produces his magic antidote and his plan of campaign against Circe.

Why does a divinity come to Odysseus' aid at this point of the story? There has been no prior council in heaven to authorize the intervention. Poseidon is still enraged with Odysseus for blinding his son and will resent any god's intervening to help the miscreant. For just this reason, her uncle's wrath, Athene withdraws her help from Odysseus from the time of Polyphemus' curse until his seventh year on Ogygia. But here is Hermes giving Odysseus material and moral support. The reason is close by. When Odysseus against all the odds, and in the face of Eurylochus' terror, takes up his sword and marches towards Circe's palace, he has once again achieved nonentity. In Polyphemus' cave he called himself No one but in going to Circe's palace, even after Hermes' warning, he has become No one indeed. He is someone whose own interests no longer concern him, who is entirely given over to the interests of others at the cost of all that he is. But the spirit abhors a vacuum even more than does nature, so that the god is irresistibly drawn to fill it. According to Meister Eckhart the miracle of virgin birth exactly symbolizes this, that god must fill every soul which has emptied itself of its own concerns.[1]

Hermes gives Odysseus the Moly plant which he seems to have pulled up on the spot:

> Then the giant-killer handed me a herb he had plucked from the ground, and showed me what it was like. It had a black root and a milk-white flower. The gods call it Moly, and it is an awkward plant to dig up, at any rate, for a mere man. But the gods, after all, can do anything.[2]

Hermes shows Odysseus the nature of the plant and Odysseus

---

1. Meister Eckhart, *Meister Eckhart*, vol. I, ed. Franz Pfeiffer, tr. C. de B. Evans (London: John M. Watkins, 1956), pp3–9, 76–77, 221.
2. *Odyssey*, x, 302–306.

passes the lesson to his hearers. Hermes shows Odysseus how the plant has a milk-white flower and a black root and Odysseus learns that Moly is the divine name for it. The plant comes with its root. It has been plucked from the ground entire, not picked, and its flower is the very opposite color from its root, milk-white against black. Only the gods can remove it from the ground and it is the sovereign prophylactic against Circe's poison. Hermes gives the plant to Odysseus but it is not at all clear what Odysseus does with it. In fact we are told absolutely nothing else about it. From this point it disappears from the story. Yet we know that this plant is his salvation from the power of the witch. He defeats her in the sign of the uprooted flower.

The emphasis here on root and flower is found in Hindu accounts of the lotus, on whose water lily flower the sage sits cross legged, resting on the points of the inturning petals. This delicate pose derives from a contemplation of the water lily as a symbol of enlightenment. The plant begins from the mud at the bottom of the pool but sends up a shoot which rises through the water until it breaks out into the upper world of the sun; here it unfolds the whole of its magnificence in the immediate presence of its lord. This is not the plant whose fruit the companions encounter after leaving the Cicones, and it is not the Moly which Hermes demonstrates to Odysseus. But the contrast between the lily's root and its flower recalls Hermes' stress on the whiteness and blackness of the Moly. And just as the eastern lotus represents the evolved soul which has reached bliss from its root in the world of illusion, so the Moly may symbolize the double nature of the human being whose base black root yields the lovely flower of compassion. But there is this important difference between the lotus and the Moly: the Moly has been removed entire from the ground, and is no longer connected to the earth at all.

This is the symbol of the human being who is liberated or enlightened while still living. The entire physical organism remains but it is no longer attached to the realm in which it developed and from which it grew. It is another instance of that switched-off fan which continues to turn through inertial force. But this is an exceptionally difficult state to achieve, as Hermes explains to Odysseus. Human beings can hardly succeed by themselves, but the gods can manage it. This is perhaps the earliest explicit reference to the need for divine grace in the human search for salvation in Western thought. It is in this sign, then, that Odysseus defeats Circe. She cannot debase him because he is uprooted and detached. As his own physical survival means nothing to him, he is dead to himself. This is the only mental temper which can resist the strong attraction to a physical self-embodiment.

Even with the Moly, Odysseus is far from confident when Hermes leaves him. In trepidation he makes the rest of his way to the palace, his thoughts dark within him. When Circe at last strikes him with her wand, he leaps into the action which Hermes has prescribed him, with an instant riposte. As Circe slips below his blade and bursts into tears, he experiences the second dazzling transformation of her repertoire. The first was the companions' shift from the polished and inlaid settles and chairs to the muddy floor of the pigsties. In this second transformation Odysseus' sword miraculously becomes his phallus:

> Put up your sword in its sheath and come with me to my bed.[1]

But though it is a trick, Hermes has told Odysseus that he must sleep with her, after making her swear the oath of the gods not to harm him nor unman him when he is naked. Given the salves which Circe possesses, it is a wise precaution not to expose the skin in her presence. But we wonder exactly what Circe would have done to Odysseus if he had gone to bed with her without having secured her oath. Would the lady's body have been so toxic that contact with it achieved what her potion had not? Is gelding what she has in mind or a mere psychological impotence? As there are two No ones in the Polyphemus story, so there are two unmannings in the Circe story: losing ones human form by being turned into a beast and losing ones virility in Circe's bed.

Soon, all too soon, Circe has sworn her oath and Odysseus stands naked before her. Sex with a goddess! He is to be elevated as far above the human level as his poor companions have been degraded

---

1. *Odyssey,* x, 333.

below it. Can he cope? It is one thing to know that the witch is on oath not to harm him nor to deprive him of his virility. It is quite another to stand before her like this, a man wasted by grief. How is he to address her? How is he to undress her? He is joining a select club, the men who have slept with goddesses, more select even than those who have visited Hades: Tithonus, Tityus, Peleus and Anchises. Will he manage this divine conjugation? He need not fear, because it was Hermes who set him up with Circe in the first place and Hermes is the god of the erection. Hermes will make sure that Odysseus holds up his head on this occasion as he ensured that Odysseus did not sink to the ground like a beast when Circe gave him her potion. Hermes is fond of mischief. He enjoys in prospect Odysseus' discomfiture as he stands to for action on that frightening bed. This is the god who jokes lustily with Apollo at the spectacle of Aphrodite and Ares in Hephaestus' trap.[1] He is the ithyphallic god whose aroused pudenda, free standing or in light relief, were on open display around the ancient city. They must have offended the Christian Arnobius as much as the phalli waved in honour of Bacchus. Here is another free-floating flower which is Hermes' gift. Odysseus enjoys a divine erection for his union with the goddess.

Hermes is the god of the vertical, of the erect human phallus and of the erect human spine. He governs the axis which stretches between the worlds from Olympus to Tartarus. His name may be connected with a word meaning a cairn of stones, and in some places where Hermes was worshipped, his votaries honoured him by each contributing a stone to a cairn. He is represented as a beautiful young man stretching upwards by the Renaissance artists Giovanni da Bologna and Botticelli. Hermes is a trickster god, a schemer like Athene, and for that reason a sympathetic ally to Odysseus, but he enjoys the embarrassing predicaments of his friends quite as much as those of his enemies. He guides the souls of the suitors down to Hades at the end of poem, since as lord of the vertical he governs both of its directions. He is here with Odysseus on Aeaea because Circe's power does not consist in metamorphosis between creatures in a single dimension, but in the embodiment of the disembodied,

1. *Odyssey*, VII, 336–342.

the shift from one order or state of being to a quite different one. This comes within the range of Hermes' own power so that he needs no Olympian authority to intervene.

After they have made love and Odysseus has been bathed in a hall of glowing gold and silver and bronze, a meal is set before him but he sits untasting. When at last Circe asks him what troubles him, he replies that no right thinking man could eat or drink while his companions were penned in pigsties; if she really expects him to eat, let her free them now so that he can see them with his eyes. It is most tactfully done, diplomacy as striking as his first speech to Nausicaa or his colloquy with Achilles among the dead. He has said nothing of this matter on his own initiative but has carefully attended to each step in the sequence which Hermes had prescribed to him. Circe asks him what troubles him and with that invitation he tells her his deepest wish. He is in no way outspoken with her though he has circumvented her twice and slept with her. He has learnt his lesson from Polyphemus.

As with the oath her response is immediate and complete. She does exactly what he asks. She drives the companions out of the sties with her wand and passes among them with ointment with which she smears them. The bristles fall off them and they are restored to their manhood but they are younger, taller and more handsome than before. They recognize Odysseus, or at least now he knows that they do, and they cling to his hands as they weep for happiness. Even the goddess is moved to pity though she did not guess why Odysseus would not eat his dinner. The imperious sorceress becomes human from this point, though we must always remember that her own burst of weeping was part of a ruse. I like best the time when she takes Odysseus aside from his companions after they have returned from the dead. Night has fallen and she sits him down on the ground and she herself lies beside him while he tells her the story of their visit to Hades.

If the companions' transformation into pigs represents or symbolizes the fall of the uncreated into a creaturely existence, then their return to the human state represents their achieving enlightenment, the state in which Odysseus himself has remained throughout the episode. They are certainly happy and such tears are commonly found in those who have achieved or are achieving spiritual realization. Like Odysseus, though enlightened they are still embodied; the root is still attached to the uprooted flower. But they have been removed from the mud in which they wallowed in the pigsties, as the plant was removed entire from the earth. The symbol of the mud or mire in which the companions wallow as pigs is another where Homer anticipates Orphic usage. For Orphics the mire was a symbol of embodiment in the physical world.

Just as Odysseus' encounter with a one-eyed giant man called Polyphemus stood for every human being's relations with their own belly, so Circe's palace is our whole world of physical nature. She is not primarily the goddess of metamorphosis; she precipitates people into their earthly embodiment. Embodiment is a state of intoxication as though the gross physical body is itself an anesthetizing torpor, an ungainly animal, a numbness in which the higher faculties are blunted or obliterated by the physical senses. This is how Homer makes us think of all that experience which is most immediate, comfortable and vivid to us. If we identify ourselves with that experience and take our own reality to consist in our physical bodies and the world of the senses, then we are truly imprisoned by our bodies, and have forgotten our homeland. This is the condition of the prisoners in Plato's cave, except that Homer's prisoners weep for what they have lost, and let Odysseus live as he struggles to save them.

Who then is Odysseus? He is a man befriended by Hermes and who sleeps with a goddess, who voluntarily enters the mortal and physical realm at great peril to himself in order to save his fellows. In this episode of the *Odyssey* Odysseus is the savior, the spiritual master who has overcome the enticements of matter and can liberate others from them. This may seem too glorious a part for our hero, but he does no less when he returns to the troubles of Ithaca from Ogygia and Phaeacia, or even perhaps when he chooses his next life in the *Myth of Er*. He does no less for anyone to whom he describes his salvation in his mystic stories across the ages. As Guénon explains it, the southern gates to the cave of the Nymphs are for immortals. But there are two exceptions:

The 'gate of the gods' cannot be an entry except in the case of voluntary descent into the manifested world, either by a being already
'delivered', or by a being representing the direct expression of a
supra-cosmic principle. But it is obvious that these exceptional cases
are not part of the 'normal' processes that we are considering here.[1]

We can see already in this first encounter with Circe the immortalized Odysseus of Ogygia and beyond. His self-sacrifice is not at
this stage a settled habit. He has been forced to it by the latest of several disasters, when he is at the very limit of his emotional endurance. I can imagine the dark thoughts which filled his heart after
meeting the ebullient Hermes. We go with him every step of his
treacherous ordeal, driven only by love.

1. Guénon, *Symbols of Sacred Science*, p160, n3.

7

THE MAGICAL SHIPS

WHAT WAS THE UNCREATED Odysseus before it became Odysseus? To what has he been restored by his self destructive ordeals? The Zen monk contemplates the question: what was my face before I was born? What was Odysseus before he became Odysseus, as Circe's pigs were companions before they became pigs? To this question we have mainly negative answers so far: Odysseus is No one, without self esteem or a sense of his own identity, without heroic values or belly or eye or his body at all. To fill the gap left by this evacuation of Odysseus we have considered Athene as the remainder or residue of the hero, on the model of the revelation made to Plotinus that his guardian spirit was a divinity. It is Athene with whom Odysseus enters the gates of the immortals on the shore of Ithaca. This was his return into the world of physical manifestation, a moment signified by Athene's removal of the mist which had concealed his beloved homeland. He has come home at last, without his companions but with the great treasure he was given by the Phaeacians, which he and Athene store in the cave of the Nymphs. He has come from Phaeacia, and this land and its people are Homer's symbol of the real self in its own place before corruption by embodiment.

Odysseus arrives in Phaeacia by a series of disembodyings and divestitures. When the storm strikes on the seventeenth day of his voyage from Ogygia he loses everything which he and Calypso had prepared. First the raft, then his clothes. In exchange for these he is given the scarf or veil by Leucothoe which helps him struggle to shore. But this, too, he must cast back into the sea after reaching shore, with eyes averted, for so Leucothoe has commanded him. Here let us take the raft and the clothes as symbols of the physical body, and the scarf as a symbol of the wraith or ghost body. With the throwing back of the scarf Odysseus discards this body, and the averting of his eyes signifies that he no longer has even the sense of sight which remains to a ghost. All that remains is the seed of the fire, in Homer's words, with which a man who has no neighbors saves himself the trouble of kindling the fire anew by burying it in the ashes. This is Homer's simile for Odysseus as he sleeps in his bed of leaves on the shore of Phaeacia.

He is awakened by the shriek of the maids as Nausicaa misthrows their ball into the river. He emerges, naked, salt encrusted, his eyes still bloodshot from their long immersion. He advances on the girls holding a branch over his privates. At the sight of him they all run away except Nausicaa. Certainly he is embodied here, but the point of the story is to realize for us as acutely as possible the embarrassing ugliness of the body from the point of view of the spirit. Odysseus is humiliated by the encounter, and Homer like Hermes takes enormous pleasure in his discomfiture. Here again our hero is stripped naked before the discerning eye of a female rigorist, though at least in this case he does not require the ministrations of the ithyphallic god to manage his part. On the contrary. This lady is far more chaste and severe than Circe and he is about as far from his best as can be imagined. Homer compares him to a lion with fire in his eyes and this may be the frightened maids' view of him. But there is a cruel and comic disjunction between Homer's splendid image and the poor man's actual condition and feelings.

Nausicaa is a tall girl, standing above her companions as Artemis does among the nymphs. She is dressed and has had time to replace her headgear between the ball's falling into the stream and Odysseus' appearance. She is the only one to stand her ground at Odysseus' approach, but as events transpire, this may have less to do with simple courage than with pity, for she is immediately and completely sympathetic to Odysseus, long before she begins to connect him with her dream husband. On Odysseus' side, then, there is a painful embarrassment and deep reluctance to intrude in his shameful state. But the princess, serene, dignified, compassionate, has the tact of her father Alcinous in smoothing any social awkwardness. This tableau represents the body abashed before the loving spirit. It is another disembodying.

But like the loss of the raft and the rest, this shame at the body is a negative symbol. It tells us what is discarded, not what is left. There are several other such representations of Phaeacia. Its court at least is personally very clean, if Nausicaa's washing of the clothes is typical. The country is at the very edge of the inhabited world; they are the last outpost of mankind. They have no enemies, partly because they are so far from the rest of humanity. More remarkable is the guarantee which Alcinous gives Odysseus when he promises to return Odysseus to Ithaca:

> And we will safeguard him on the way from any further hardship or accident till he sets foot in his own land. After which he must suffer whatever Destiny and the relentless Fates spun for him with the first thread of life when he came from his mother's womb.[1]

It appears that Phaeacia can protect Odysseus beyond the reach even of the Fates, so that Odysseus' own destiny is, as it were, in remission while he is travelling on the Phaeacian ship. But though this may hold for Odysseus it does not seem to be true of Phaeacia itself whose doom at the hand of Poseidon has been long foretold.

The two positive symbols of the beyond in Homer's description of Phaeacia are the garden of Alcinous and the magic ships. In the garden all the plants are at their different stages of development from seedling to crop simultaneously, and the west wind blows gently all the time. Homer's description of Alcinous' garden was quoted in the first chapter in the context of astrogeography. Diodorus Siculus quotes from Homer's description of Phaeacia at this point, and compares Phaeacia to Taprobane or Sri Lanka, which is six degrees north of the equator. Diodorus was writing in the first century BC. But the garden of Alcinous must be understood together with the magical ships as the two halves of a single idea. These ships go anywhere on earth and back again in the course of a single day. They seem to be able to circumnavigate the globe at the speed of the stars. They are swifter than a falcon on the wing, as swift as thought. They need no pilots for they have minds of their own, but must be told where they are to go. The Phaeacian crew provides the motive power

---

1. *Odyssey*, VII, 195–198.

by rowing and they are clearly the fastest rowers on earth. The ship which takes Odysseus home is on its maiden voyage and is turned to stone and fixed to the sea bed on its return.

In the sixth century BC the poet and philosopher Parmenides composed the first Western ontology or theory of being. According to Parmenides reality is rather different from how we suppose it. We suppose the world to exist materially and to undergo change through time. But in fact reality is atemporal or outside time:

> Neither was it ever, nor will it be, since it is now altogether, one and continuous.[1]

But though reality is atemporal, this does not mean that it is, as it were, empty of time. On the contrary it is overflowingly full of time like the garden of Alcinous. For in reality all the different moments of time of which each of us has knowledge are not a sequence in one of which we are presently, in one only and no other. In reality all these moments are present simultaneously to the mind and it is only the poor body which has to drag itself from one to the next, instead of flying like the mind wherever it will. Each of us thinks of the self as standing at a point on the surface of the earth which stretches out around us in all directions from that point. Just as we think of ourselves at one moment of time in an indefinitely long sequence of moments, so we think of ourselves as being at one place at any time among an indefinite number of places. Reality, Parmenides abjures us, is different:

> But, look, whatever is absent is firmly present to the mind nonetheless. For you will not cut off being from holding on to being, neither by scattering it everywhere in all directions in order, nor by putting it together.[2]

The entire physical universe and all of its changes in the process of time, so far as we know them, are present simultaneously to the mind, however bound the body may be in its apprehensions to its immediate circumstances.

Parmenides' ontology is as much a meditation on the magical

1. *Ancilla to the Presocratic Philosophers*, op. cit., Parmenides, frag. 8.
2. Ibid., frag. 4.

powers of Homer's Phaeacia as Empedocles' *Purifications* are a meditation on the Cyclops episode and Plato's *Myth of Er* is a reconstruction of the Circean books. Parmenides has pondered Homer's description of the Phaeacian ships which are swift as thought. For Parmenides the powers of the mind are the foundation or core of the human personality, not the body. Nor can the mind be a product of the body in the traditional view, since it is impossible for more to emerge from less. The human mind must have come down to us from heaven. The human being is not a complexification of biological life which is itself a complexification of physical or chemical process. This is impossible. The human being is a crudification or grossification of a spiritual principle, a god reduced or covered over. Reality is eternal, omnipresent and all knowing and the human mind is much closer to possessing these powers than is the human body. Phaeacia is the symbol of the human mind before its fall into embodiment.

This mind or *nous* is the seed of fire which is, as it were, implanted in the human body and which organizes and enables experience. In Phaeacia Odysseus is in the kingdom of the mind and here he relates the tale of his adventures after leaving Troy. His stories are largely symbolical accounts of the body as understood from the point of view of the mind: of the voice, the belly, the eye, and the body as a whole in the Circe episode. In the kingdom of the mind Odysseus sees the body from the point of view of the mind so that his stories are exceptionally well fitted to the Phaeacian temperament and the Phaeacians are very receptive of them. The Odysseus who is carried back to Ithaca on one of their magical ships is this same pure mind on its way to embodiment or re-embodiment in the cave of the Nymphs. On leaving Phaeacia he falls into a sleep delicious and profound as soon as the oars of the Phaeacian ship touch the water. The souls in the *Myth of Er*, after choosing their lives and crossing a treeless plain, are required to drink of the river Unmindfulness whose waters no vessel can contain. Then these souls, too, fall into a deep sleep and in the middle of the night there is thunder and lightning and the souls rush up in various directions like shooting stars to be born. But these souls when they awaken in this world have no recollection of that other world from which they came here. They may with difficulty recover the memory of it if they drank no more than they had to drink of the river. But Odysseus returns to Ithaca in full and conscious memory of Phaeacia when he wakes up in Ithaca. His pure mind, his real nature, are intact.

Alcinous has told Odysseus that while he is on the Phaeacian ship his destiny is suspended. As soon as he returns to Ithaca, the thread of his destiny will resume. While with the Phaeacian ship he is beyond the power of the Fates, or Spinners as Alcinous calls them. Similarly the souls in the *Myth of Er* who are choosing their next lives are beyond destiny at the moment when they freely choose. This is symbolized by their being beyond the vast complex of Necessity's spindle and the bands of the fixed stars and planets. Their destiny resumes at the moment after they rush up like shooting stars to be born. They are like shooting stars because at that moment they cross all the astral and planetary bands on their way from beyond the heavens to the earth, traversing the whole of space. They rush up and not down as shooting stars appear to us. This emphasizes the vertiginous nature of the transition between the worlds. In the same way the magical ship of the Phaeacians transports the sleeping Odysseus across the astral and planetary courses and back into the world of physical manifestation. The Spinners described by Alcinous are those same goddesses whom Plato describes: Necessity, Lachesis, Clotho and Atropos. Alcinous' word for his spinners is *Clothes*. The Spinners are also the whirlpool Charybdis, the six heads of Scylla, the singing Sirens, and the Cattle of the Sun. Odysseus has escaped them once, and now re-crosses their courses on the Phaeacian ship at perfect rest. So if Phaeacia is on the equator, it is clearly on the celestial equator.

We may compare the relations between the omnipresent mind and all the entities which comprise the physical world to the relations between the center of a sphere and all the points on its surface. All these points on the surface are equidistant from the center in the great and final equality of creation. As Parmenides says:

Neither is anything less but everything is full of being.[1]

---

1. *Ancilla to the Presocratic Philosophers,* op. cit., Parmenides, frag. 8.

I have already quoted Parmenides' use of the symbol of sphere and center in my account of the Aeolus episode. For Parmenides being is:

> like the mass of a well rounded sphere stretching equally in all directions from the center.[1]

The center of being, the point from which it all emerged, is that first point symbolized by the bag of winds which are also the directions. It is symbolized also by the floating island of Aeolia, its self containedness, piety and peace. Aeolia is the other point from which, as from Phaeacia, Odysseus is returned to Ithaca. But Phaeacia is on the very edge of the inhabited world, an edge which corresponded more or less to the equator according to ancient geographers. How then could it be a center like Aeolia? It is a center in effect because nowhere on the inhabited earth is more than a day's journey with return from Phaeacia, and there is no place on earth closer on average to the rest of it than this.

But the relations of mind to physical things are not quite the relations of the first point to the created world. We might say that the power which generates the world realizes itself in us who are in the world in the form of our minds. Our minds are more like that power than anything else in nature. This power brings the world into existence by expanding as a sphere, each point on the surface of which is joined to the center from which it began by a unique radius. The point on the surface is sustained in its position by being one term of this radius, of which the other term is the center of the sphere. So each creature in the world is the embodiment of a unique ray from the source of all, and since that ray is unique to the creature it cannot be duplicated or repeated. This is why Aeolus can do nothing more for Odysseus when he returns to Aeolia after nearly reaching Ithaca, and why the Phaeacian boat which takes Odysseus home is on its maiden voyage and is destroyed on its return.

I have imported the sphere from Parmenides to explain the symbols of Aeolia and Phaeacia. But there are spheres enough in Phaeacia. There are two, the ball with which Nausicaa and the maids are

---

1. *Ancilla to the Presocratic Philosophers*, op. cit., Parmenides, frag. 1.

playing as they sing and the beautiful purple ball, made by Polybus, with which Halius and Laodamas play as they dance.[1] The Phaeacians excel at these games as they do not at the martial arts: nor need they since no enemies come near them. Nausicaa's ball falls into the river; the maids scream; Odysseus awakes. Is this not a symbol of awakening; a ball falling into the stream of time from the multidirectional play of dreams?

* * *

There are over twenty-eight thousand hexameter lines in the *Iliad* and *Odyssey*. This is a vast area to cover even at the most superficial level. But the situation is far worse: wherever we lower a line to take a sounding, the lead slips down through the water, the deeps open themselves to receive it and it never reaches bottom. However far and however successfully the enquiry proceeds, there is always more and more. In this ocean we have been investigating the books of the *Odyssey* which relate Odysseus' adventures on his journey. These are the veritable Bermuda triangle of the Western tradition, into which over the millennia innumerable craft have ventured, to disappear without trace. After thirty years of considering and teaching this part of the *Odyssey*, I say with confidence that Homer is impossible. The line which will plumb Homer has no end. And, of course, the fact that the task of understanding Homer is impossible does not diminish in the slightest our duty to undertake it.

There is an apocryphal story describing the first time Jesus went to school.[2] Sitting there he watched carefully as the teacher wrote the letter aleph in front of the class, and he complied with the teacher's order to write it on his own slate. Then the teacher wrote down the letter beth and told them to write this letter on their slates. At this point Jesus caught the teacher's attention and said to him 'You have not explained aleph. How then can you go on to

1. *Odyssey*, VIII, 370–379.
2. *The Apocryphal New Testament*, tr. and ed. J.K. Elliot (Oxford, Clarendon Press, 1993), 'The Infancy Gospel of Thomas' and 'The Gospel of Psuedo-Matthew', pp76–77, 81–82, 90.

beth?' This drove the teacher into a rage with extraordinary consequences, but Jesus had a good question. We are all, ancestors and
contemporaries alike, faced with the same question by the miraculous beginning of the epoch which we share. How can we proceed
to an understanding of our civilization before we understand that
great bolus of verse at its start, the *Iliad* and *Odyssey*? We cannot,
but we cannot understand these poems either.

This suggests a quite different model of civilization from the pro-
gressive one which we have recently adopted. On this alternative
vision civilizations come into existence all at once, as at the wave of a
magic wand, entire and perfect in every part and with no antecedent
development which we may discern. Homer, the Vedas, the Torah, I
Ching and the Book of Songs, these supreme works appear at the
fountainheads of their civilizations and each of them imposes an
insoluble riddle on all succeeding generations. The canonic works of
a tradition are commentaries on the masterwork from which the
tradition began. We have seen how this is true of Parmenides' and
Empedocles' poems and of Plato's *Myth of Er*. Even the Acts of the
Apostles derives in some sense from the *Odyssey*. In this way the
course of a traditional civilization is not outwards towards some
hitherto unachieved equilibrium; it is an unfolding of the possibilities in some original kernel, seed or bud. There is always a clear sense
of how the present world relates to that primordial work. That work
itself is regarded as impossibly wise, beyond all finding out. This
was, for the most part, the Greek and Roman view of Homer.

The unfolding of Homer in the ancient Greek world is commonly
and correctly described as the transiting from a mythical to a logical
understanding, from the stories of Homer to the dialectical demon-
strations of Socrates as Plato presents him. The aim becomes to fix
the terms of discussion and their interrelations firmly in one's mind,
so that in their deployment the argument is won. This fixing of
them is called conceptual analysis and it is hard to know how much
is gained by it, apart from the winning of arguments. Plato, for
example, develops the dialectical account of the soul in the *Phaedo*
and the *Republic*: ever since, the term has become part of the philosophical lexicon. But does this analysis of the concept of the soul
assist us on the spiritual path, or does it lead to sterile confusion? In

Homer's work the term soul is confined to the wraiths of the dead
and he has no term for what makes Odysseus immortal. He indicates what he means by metaphor and analogy at considerable
length in unforgettable stories, and it is arguable that the stories are
far clearer than the analysis, for all that it is finally impossible to sys-
temize them. I do not mean that the stories are better teachers than
the analysis. I mean that the stories actually convey a far better picture or account of these realities than the analysis of the dialectic
ever could. This was not a problem for Plato, since he and his readers had both the analysis and the stories, but it is certainly a problem
for those ages in which logical demonstration has displaced the
mythic in the search for wisdom.

The Archaic Greek world, Homer's world, has much in common
with the Romanesque period of European history. The Archaic was
at least as different from the Classical in Greece, as the Romanesque
from the Renaissance. In both Archaic and Romanesque there is an
identity of interest between the priest and the peasant, in which the
forms of religious teaching are fitted to the agrarian life. In both
Archaic and Romanesque, religious education was accomplished
almost entirely through the medium of story and image rather than
through theology or philosophy. Story may well be the optimal
means for the developing of the spiritual sense, both because it represents spiritual truth more accurately than analysis can and
because it is so widely appealing.

The Archaic period is the spiritual highpoint of the Greek civilization as the Romanesque was of Christendom, and both the
Archaic and the Romanesque decline into rationalization and logic-
chopping. The dialectic of Platonism has its counterpart in the
scholasticism of the Middle Ages. To the logical it no doubt appears
that the poet or the Evangelist is in their chains, though Plato was
saved from the worst excesses of this by the memory of his own
youthful ambition to be a poet and by his abiding love of making
myths. These facets of his character may explain Plato's extraordinary ambivalence to Homer, especially in the *Republic*, where that
homage to Homer, the *Myth of Er*, completes a dialogue in which
Homer and the poets are excluded, on Socrates' account, from the
ideal state. But we must remember here Plato's liking for subjecting

even his own philosophy to the most critical attack, as when his fictional Parmenides dismembers his theory of ideas. In the *Phaedo* the condemned Socrates is busy turning the stories of Aesop into verse in order to meet the Oracle's command that he devote himself to the Muses[1]. Plato's aim is not to persuade us of his doctrines so much as to make us think about the issues on which they touch. In his critique of Homer's anthropomorphic pantheon, Plato is actively sustaining the spiritual purity of his tradition, not despite Homer but along with him against his vulgarizers. Such was the view of the later Platonists.

A similarly uneasy relationship between the symbolic on the one hand and the logical on the other characterizes the latest ages of Christianity. The simple story of the Evangelist is pitted against the unbelief of the fact-hunting materialist. Here the most profound symbols, the Immaculate Conception or the Resurrection, become stumbling blocks to faith. But it is their very unlikeness to the normal order of things which gives these events their immense power to concentrate and illuminate the mind of the student. These are facts but they are facts of a quite different order. Meanwhile the Homeric poems have almost passed beyond the range of the religious scholar, the philosopher, the astronomer and the historian of science. Their poetic and narrative form prevents our understanding them as the ancients or as Thomas Taylor did. The *Iliad* and the *Odyssey* have become works of literature, where they are rated as highly as any, but they are not considered in the terms in which we have considered them here.

In our time the *Odyssey* is saved for each new generation in its new incarnation as a literary classic, and at this level its appeal has never been greater. The E. V. Rieu Penguin translation of the *Odyssey* is one of the very best selling books of all time. But here and there will be a student who resonates to the poem in ways which teachers and secondary reading do not explain. I know for I was one. I came upon Taylor's translations of Porphyry and Proclus on Homer and the real work began. I do not know how this book of

---

1. Plato, *Plato. Collected Dialogues*, eds. E. Hamilton and H. Cairns (London: Loeb Classical Library, 1961), 'Phaedo', 60c-61b.

Taylor's essays found its way to me. I could find no explanation. Nor do I accept Taylor's own readings of Odysseus' adventures.[1] But Taylor's passion to understand Homer as a sage kindled me. My accidental opportunity to reconsider Homer may not be given to all the students who have a genius for this thinking since Taylor's work is very little known.

The delimitation of Homer as a literary artist has other disadvantages. Not to read Homer as a religious teacher, nor as a scientist, is to refuse to acknowledge the claims made for him by his tradition. These claims must at least be assessed. They cannot be dismissed on principle, however hard, even impossible, the task of assessing them may be. To be sure, we do not live in a religious community in which the daily forms of life derive directly from the Homeric teaching. For us Homer is the beginning of the epoch in this very special sense, that he has no context against which we can evaluate him. He is his own context because we know very little else about his time. But Homer is also self-contained. I have compared the *Odyssey* to a puzzle or game in which we assume that all the moves and pieces are to hand. It is a game full of tricks and turns like Odysseus himself, and we are to find a path which links its various episodes in a single programme.

Not to try to see the *Iliad* and the *Odyssey* as the works of a religious teacher, a philosopher and a scientist is to fail in hope. Classical scholars should demand more from the studies to which they devote their lives. In particular, if we separate Homer's poems from the scientific and philosophical developments which followed, we at one stroke remove our greatest aid to understanding Homer and we cripple our understanding of early Greek thought after Homer. By postulating that Homer was a scientist and a religious teacher, I have been led to draw freely on the works of the Presocratics and of Plato, seeking connections between their concerns and those of the *Odyssey*. In the case of the astronomy of Book XII, I would not have seen the book as astronomical if I had not read Plato's *Myth of Er*. Empedocles' *Purifications* and the poem of Parmenides both illumi-

---

1. *Thomas Taylor the Platonist*, op. cit., note 13, Porphyry's 'Concerning the Cave of the Nymphs', pp322ff.

nate Homer and are illuminated by him. I understand the *Purifications* more viscerally from those horrible moments in the Cyclops' cave, and the Phaeacian ships and the bag of winds brighten my contemplation of Parmenides' logic in the Way of Truth. But as with Plato, so with Porphyry: I would never have guessed the astronomical significance of the gates to the cave of the Nymphs, however hard I had studied the *Odyssey* by itself.

What have Pythagoreanism and Orphism to do with Homeric religion? Herodotus tells us that what was called Orphic and Bacchic was really Pythagorean and Egyptian.[1] But Herodotus also tells us that the Greek gods are Egyptian except for Poseidon and the Dioscuri, and that Homer and Hesiod introduced the names, functions and forms of worship of all the gods into Greece. For Herodotus, Orphism and Pythagoreanism and the Homeric religion all came from Egypt. The adventures of Odysseus present us with an horrific vision of ourselves as carnivores in the Cyclops episode. In the Circe episode humans are represented as trapped in the bodies of pigs, mired with the mud of their sties. In Book XII are images of astronomy and music. There is little in Pythagoreanism and Orphism which may not be found in full or in embryo in Homer, and the Orphic and Pythagorean philosophers are our best guides to the ancient understanding of Homer. Orphics and Pythagoreans were, perhaps, two kinds of Homeric fundamentalists; more austere, intellectual and passional devotees to the teachings of the poet; an Homeric elite like the Sunyasin of Hinduism or the Sufis of Islam. From our place in time, Homer needs the philosophers and the philosophers need Homer if we are to understand either.

* * *

Finally, another poem: *Mythistorema X* by the modern Greek poet George Seferis:

> Our country is closed in, all mountains
> that have the low sky for a roof day and night.
> We have no rivers, we have no wells, we have no springs,

---

1. Herodotus, op. cit., II, 132.

Only a few cisterns, and these empty that echo and we
worship them
A stagnant, hollow sound, the same as our loneliness
the same as our love, the same as our bodies.
We find it strange that once we were able to build
our houses, huts and sheepfolds.
And our marriages, the cool coronals, and the fingers
become enigmas inexplicable to our soul.
How were our children born, how did they grow strong?

Our country is closed in. The two black Symplegades
close it in. When we go down
to the harbours on Sunday to breathe
we see, lit in the sunset,
the broken planks of voyages that never ended,
bodies that no longer know how to love.[1]

Seferis was an admirer of T.S. Eliot and this poem is a modern
Greek Wasteland. But Seferis' use of desertification is closer to the
Greek landscape than was Eliot's to the urban canyons. Seferis'
hollow cisterns and their echo which is the same as our bodies give
new force to Eliot's Hollow Men, who were in any case stuffed with
straw. But Seferis' most chilling lines declare wonder at the building
of the houses, huts and sheepfolds once upon a time. This wonder
suggests a people in the last stages of destitution and apathy, to
whom even the simplest and crudest effort seems beyond their
capacities. On Sundays, at least, they breathe, but in the waters of
the harbour they see broken planks, images of the despair from
which they are escaping.

It was not always so. They or their ancestors did build those build-
ings once. Their marriages have become enigmas; they were not
always so. Once they knew how their children were born and grew
strong, and their bodies knew how to love. How were things with
them then? Were they covered over and hemmed in by mountains?
Did the black clashing rocks close them off? But whatever they were,
they have sunk to the point where that now seems miraculous in
retrospect. This is recognizable as Eliot's modern world, a state of

1. George Seferis, *Collected Poems 1924–1955*, tr. and ed. by Edmund Keeley and
Phillip Sherrard (London: Jonathon Cape, 1973), 'Mythistorema X'.

lethargy surrounded by a meaningless welter of wealth or poverty. We join a Greek Sunday afternoon in the last stanza, but one from which all the delight has been stripped.

Perhaps it has always been so, and every human being, every human town, is a closed country, isolated in a small pocket of the world and of history, ignorant of how it got there or what will happen when it ends. Things always made more sense in the past than they do in the present, and every age is struck by the wisdom and skill of its predecessors. Seferis is not painting the modern condition but the human condition in general. It is, we suspect, a peculiarity of humans at all times that they want to know the answers to unanswerable questions. The other animals either do not trouble themselves with the questions or have long ago found the answers. In either case they are unconscious of the limitation which confines the singers of Seferis' song. In this song the whole of the human race may join with sincerity. We are all closed off. Everyone has always been closed off.

In fact, of course, the singers of Seferis' chorus are describing not the modern, nor the general, but an ancient predicament, the predicament of the Phaeacians after the return of Odysseus to Ithaca. Here are the famous Phaeacian singers again and we may rejoice that whatever their sufferings, their powers of song are unimpaired. They are singing of the doom that was long predestined to overtake them at the hands of Poseidon in his rage at the help which they had given to Odysseus by returning him to Ithaca. As the Phaeacian ship which has left Odysseus in Ithaca nears the shore of Phaeacia, Poseidon strikes it with the flat of his hand and in an instant it petrifies and fastens to the seabed. The whole Phaeacian people are watching the ship as it approaches them, just as the whole people see the broken planks of voyages which never ended on their Sunday promenade in Seferis' poem. Broken planks are no substitute for Homer's rock but they answer well enough to Odysseus' stormy shipwrecks off Thrinacie and Phaeacia, if never to the experience of the Phaeacians.

But the doom which had been told to the Phaeacians by Nausithous, Alcinous' father, went much further. Not only were one ship and its crew to be wrecked, but Poseidon would throw a great mountain or mountain range around the Phaeacian city. The turning to stone of the ship, the surrounding of the city by mountains, these are the works of the Earth Shaker who is also the earth shaper. What happened to the ship we know, but what happened to the city we do not know, for though Zeus suggests to Poseidon that he ring Phaeacia with mountains, and though Alcinous exhorts the Phaeacians to placate Poseidon so as to prevent the earthquake, Homer cuts back to Odysseus on the shore of Ithaca at the very moment when this event will or will not take place. But Seferis has no doubt. His Phaeacians are closed in by mountains, and by the black clashing rocks which Circe mentions to Odysseus in her account of his journey from Aeaea to Thrinacie. The Phaeacians still have access to their harbour and the sea but these things too remind them only of what they have lost.

Two events and one possible event according to Homer: the turning of the Phaeacian ship to stone; the re-incorporation of Odysseus on Ithaca; and, possibly, the surrounding of Phaeacia by mountains. These events occur more or less simultaneously in Phaeacia and Ithaca which are widely separate but connected by the magic ship. First, the petrification of the ship; then the debate and the awakening. After awakening, Odysseus meets Athene with whom he enters the cave of the Nymphs where he is symbolically reincorporated. At more or less the same time as this, the ship which brought him is turned to stone and Phaeacia is about to be hemmed in by mountains. These also are symbols of incorporation. As Odysseus is re-embodied in the cave which is the womb, so what he had been in Phaeacia, the seed of fire, the mind in its own place beyond the Fates, is powerfully transformed in parallel. The events in Phaeacia represent as directly as possible how the subtle powers which pre-exist the physical are changed into the being manifested.

We must add the lives of the fifty-two crew on the Phaeacian ship to the tally of damage done by Odysseus' return. Theirs is a strange death. The ship left Phaeacia soon after dusk and arrives at Ithaca a little before dawn. But it has already returned and been petrified by the time Odysseus awakes on Ithaca with most of the day before him. The explanation is likely to be that on the outward journey, the weight of the Phaeacian guest gifts slowed the ship, which reached its full speed only on its return. There on the Phaeacian shore the people awaited it, their maiden ship crewed by their finest crew, nearing home and unencumbered. Those rowers were making a special effort in their last few moments. Then, instantaneously the ship is stone, the rowers barely thrown forward on their benches by the dead stop at that speed. Of all the images of incarnation in the poem, and we seem to have seen a score, this is the most amazing. In a flash the magical omnipresence of the mind is fixed to a single spot and turned to stone. There is no process here, no stages of embryonic transition. The omnipresent and atemporal spirit enters within the world as an individual among other individuals. It enters suddenly and completely as a fly dies squashed on a table top.

The mountains which may surround and conceal Phaeacia are also symbols of individuation. This country which was temporally and spatially universal by virtue of its garden and its ships, is now completely isolated from the rest of the world. In the same way the spirit or mind, which has been one in all now becomes one among others and divided from them. Meanwhile Odysseus is being made anew from those marvellous fabrics of sea purple on his loom of stone. The mountains which may enclose Phaeacia, like the loom, are bones, the bones of the cranium which surround the brain, with an opening at the top. The human head, according to Plato, is the first organ to be devised in the plan of creation.[1] Here we are given an account of how its carapace was formed. Phaeacia is sealed off all round; its ships, they swear, will never carry strangers to their homes again. Phaeacia is lost and Odysseus never knows it, any more than he knows of the deaths of the crew who brought him home. But he remembers Phaeacia and tells Penelope of his experiences there.

And Phaeacia is lost to us, though no more lost to us than it was to Homer and Odysseus since they still tell us of it. In the end however the *Odyssey* is not the story of apotheosis, but is the story of anthropogenesis. In the story Odysseus becomes an immortal and returns to earth, but Homer is less concerned with the beyond than with the organs of our human bodies as seen from the point of view

1. Plato, *Timaeus,* op. cit., 44.

of the spirit. The breast, the breath and the voice; the belly and eye; the bristles and spine; the phallus and womb; the mind and skull: the formation and nature of these are the poet's chief concern. To this we should add our place within the whirling orbits of the stars and our relations to those husks of people in the afterworld. Homer is not the only teacher of these things in the Western tradition but he is the earliest, the most complete and by far the most brilliant. Not to know and understand Homer in this way is to suffer the fate of the singers in Seferis' chorus. To them marriage has become an enigma, the birth and growth of their children incomprehensible. Their bodies at last have lost the knowledge of how to love. Ignorant of how they came to be, they are surrounded by the debris of the civilization which once could have told them.

We no longer know what we are, as Odysseus knew it in Phaeacia and as Homer knew it. This knowledge is not edifying in the ordinary way. To see oneself as a cannibal Cyclops or as a pig in Circe's sties does not increase one's sense of human worth. Homer, on my view, was a world hater and a life denier. I agree that he also celebrated the human world as passionately and as variously as anyone in our epoch. This is a peculiarity of renunciates, that they have life more abundantly.

PARMENIDES AT DELPHI

Rule, compasses and square
Are quite precise, Cyrnus,
But any man to whom
The Delphic priestess gives
An oracle from God
Must be much truer still.

There's no escape for you
If you add anything
To what the priestess said;
If anything's left out
You can't escape your sin
Within the sight of God.

—Theognis

INTRODUCTION

We may accordingly define a metaphysical sentence as a sentence which purports to express a genuine proposition, but does, in fact, express neither a tautology nor an empirical hypothesis. And as tautologies and empirical hypotheses form the entire class of significant propositions, we are justified in concluding that all metaphysical assertions are nonsensical.[1]

THROUGHOUT our epoch and around the globe many people have discussed metaphysical matters. There is even a Chair of Metaphysics in the University of Oxford. But Ayer says metaphysics is nonsense. For this empiricist the nearly universal testimony of humanity is worth nothing. Tao, Brahman, Parmenidean Being, Jehovah who says 'I am I am': it is all rubbish, the product of errors in logic which can be clearly exposed by two or three pages of analysis.

Such 'analysis' is the mainstay of philosophy as it is practised now in the English-speaking world. No doubt very few philosophers in this movement would presently endorse Ayer's arguments in every part. But few philosophers now would disagree with Ayer's view in principle, that ancient and medieval metaphysics is largely bunkum. Bertrand Russell represented Western metaphysics in his *History* in the same way. We may trace this same approach back to John Locke, at one time lecturer in Greek at Oxford, who believed that the medieval Scholastics had created a mass of philosophical confusions which his quite novel analysis would resolve.

What Locke dared at the beginning of the Eighteenth century, Ayer and Russell inherited. The middle years of the Twentieth century, when Ayer and Russell were writing, were the period of the greatest persecutions in our epoch, and the religious particularly were victims. The atheist empires of Hitler, Stalin, and Mao demoted these people to the subhuman and removed them. Given this historical context, the total rejection of all traditional metaphysics and theology

1. A. J. Ayer, *Language, Truth and Logic* (Middlesex, Penguin Books, 1936), p24.

as pure nonsense suggests a closer affinity between the Anglo, Nazi and Communist mentalities than is usually acknowledged. Modernity triumphs, in any guise. Religious people are not wrong, exactly. They are so stupid, they cannot see how all their spiritual talk is just jabber and gesticulation. They are on the other side of the glass.

Certainly Ayer cannot hear them there. He is deaf to the language of metaphysics and theology and has no understanding of its rules. Consider the following passage:

> But in that case the term 'god' is a metaphysical term. And if 'god' is a metaphysical term, then it cannot be probable that a god exists. For to say that 'God exists' is to make a metaphysical utterance which cannot be either true or false. And by the same criterion, no sentence which purports to describe the nature of a transcendent god can possess any literal significance.[1]

Among those with very little knowledge of metaphysics, the question 'Does God exist?' is a live one. But metaphysicians and theologians suppose this very question to be hopelessly misconceived. From their point of view, too, God most certainly cannot be said to exist, since existence is a mode of reality which depends upon or is contingent upon what does not exist but truly is. Metaphysicians suppose that there is a difference in kind between things that merely exist and God: all things exist by virtue of God who is their ground and principle; but God is God entirely by virtue of God's own nature and is neither dependent nor contingent upon anything else at all. This is a tautology of metaphysics, a matter of definition. God is; things exist. But this elementary distinction between divine and contingent modes of reality makes the claim that God exists an embarrassing logical error in metaphysics, let alone for Ayer. Ayer could hardly have framed a clumsier example of a metaphysical proposition about god than 'God exists', since no metaphysician would accept it. On this matter Ayer is plainly ignorant of the metaphysics and theology he dismisses briefly and finally. He cannot hear them, though he took his degree at Oxford, one of the theological foundations of Europe.

1. Ayer, op. cit., p120.

Since Ayer and Russell, the philosophical temper of the Anglo-sphere has not changed. It is still as far from traditional metaphysics as it can get. The tone-deaf are still running the music school, and doing the teaching and grading. So we should not expect much from recent accounts of Parmenides' thought since Parmenides was the first metaphysician in the West to make this distinction between what *is* in the fullest sense and what merely seems. Parmenides' distinction was taken up by Plato for whom it differentiated the world of eternal, unchanging and creative ideas from the world of things which come into existence and then cease to exist. Parmenides is the fountainhead of metaphysics in the Greco-Roman tradition, so his account of Being is now explained away as one or another logical error after the manner of Ayer and Russell. Poor Parmenides! Poor Westerners to have believed him for so long!

Not only the metaphysics but Parmenides' historical context, too, has been underworked in recent accounts. Herodotus' story of how Parmenides' city was founded turns on a reinterpreted Delphic oracle; much of Parmenides' poem is delivered in an oracular style in oracular hexameters. What has the poem to do with the events described by Herodotus, events which occurred at or shortly before the time of Parmenides' own birth? Peter Kingsley is one of the few scholars to have raised these matters recently.[1] Consideration of them here brings into focus certain literary remains of Parmenides' time, including two poems about Parmenides' poem and addressed to Parmenides by a contemporary.

Finally there is Parmenides the natural scientist. Here we have little to guide us. One fragment promises an account of how the universe came into existence 'as far as the ultimate Olympus'; another fragment describes a gestatory syndrome leading to hermaphroditism. There may be a connection between the cosmogony and the embryology but what is it? And are either or both related to the tradition that Parmenides was the first to divide the terrestrial globe into five zones, including two polar zones and one equatorial? Then there is Plutarch's high praise of Parmenidean natural science.

1. Peter Kingsley, *In the Dark Places of Wisdom* and *Reality* (Inverness, CA: Golden Sufi Center, 1999, 2003).

Plutarch wrote these words six centuries later, separated from Parmenides by the findings of Eratosthenes and Aristarchus:

> Parmenides has much to say about earth, heaven, sun, moon and stars and he has recounted the genesis of man. And for an ancient natural philosopher who has put together a book of his own, and not pulled apart the book of another, he has left nothing of real importance unsaid.[1]

---

1. Plutarch, *Moralia*, vol. xiv, tr. F.C. Babbit, et al. (Cambridge: Harvard University Press, 1936), 'Against Colotes', 1114B.

The Prologue of Parmenides

The horses that take me as far as desire
Were escorting me when they led me and came
To a road whereon there were many speaking
To the road of a Demon Goddess, taking
Through every city the man who knows.

There was I taken, the fillies all talking
Were taking me there as they strained at the car
And maidens they were who were leading the way.

The axlepost like a pipe was ashrieking
Ablaze in the boxes, awhirl on the wheels
Whenever those maidens, the Daughters of Sun
Would speed their escorting, after first leaving
The Houses of Night for the light and pushing
Away from their faces their veils with their hands.

There are the gates of the paths of Night and Day
Around them a lintel, a stone sill beneath
High are they as Heaven and filled by great doors
And their changing keys cruel Justice keeps.

Her the maidens appeased and gently speaking
Knew well to persuade her to unwing the doors
And push out from the gates the rounded oak bar.

Yawned wide, then, the chasm they made of the doors
Folding back in their pipes, revolving in turn
One after the other the doorposts of bronze
Bestudded with bolts and pins.

There straight through them
The maidens took horses and car down the road
And a Goddess, waiting, received me and took
My right hand in her hand, and spake thus to me

'Young man, companion of deathless charioteers
These horses that bring you as far as our House,
Welcome!'

1

ELEA

MOST OF WHAT I SAY in this book I have learnt from an ancient Greek poem, or rather from the fragments of the poem which have survived to us. The chapters after this one interpret these fragments. This chapter is concerned with the life of the poem's author, Parmenides, citizen of Elea, son of Pyres. Parmenides lived at some time between 540 and 450 BC and is still honoured as a founder of western metaphysics, to which he is reckoned to have contributed the notion of being. His account of being is only a part of the part of his poem which has survived to us. A large part of the fragments is natural science and this is why this poem, like many others, was traditionally called Concerning Nature. This part of his poem is little known or understood. When it is understood, we see that Parmenides was also a great poet through whom, as through Homer, there comes to us in a perfected form an ancient priestly and bardic tradition. Great poet, founding metaphysician, astronomer, psychologist and biologist, Parmenides was more. He is also said to have made laws for his city Elea. It would be hard to find in the histories of ancient times a genius of more diverse attainments or one whose work has had a more lasting success. Yet much of Parmenides' work has been ignored or misunderstood and his fame should be greater than it is.

The life of a person is largely the life of his people and this is true of Parmenides as of anyone. He was born, we are told, to a rich and illustrious family which at about the time of his birth suffered a series of catastrophes. These events must have impressed him deeply for they impressed the whole of Greece. They were recorded in detail a century later by the historian Herodotus and his account has survived.[1] We hear also that Xenophanes, a teacher of Parmenides, wrote an epic poem which told the same story.[2] His poem has not survived but it is clear from Herodotus' account why the events may have engaged an epic poet. It is a story of heroes.

The generation of Pyres, Parmenides' father, was not born in Elea in southern Italy, the city for which Parmenides made laws. At that time Elea did not exist. The generation of Pyres was born Phocaean, in the city of Phocaea many hundreds of miles east of Italy on the most western coast of Asia Minor near modern Izmir. Phocaea was one of the cities founded at the time of a great eastward expansion of the Greeks when they occupied the Aegean islands and the western coast of what is now Turkey. This happened about 1000 BC. The emigrants started from Athens under the divine leadership of Delphic Apollo and the Athenian hero Ion. They founded their colonies in geographical groups, each claiming for itself distinct ancestral origins and forming a federation. On and off the coast of Asia Minor there were the Aeolians to the north, the Ionians in the center, the Dorians to the south. Their ancestral origins were of the greatest importance to these settlers of the new eastern lands. Herodotus derides the Ionians there of his own time for pretending to be of the purest blood among the Greeks. Phocaea was accounted Ionian, the northernmost city of the Ionian federation and one of its leading members. But there was a tradition that the land of Phocis was its ancestral home on the mainland of Greece, not Ionian Athens, and that the name of Phocaea was derived from Phocis. Phocis is the country in central Greece where Delphi is situated, the oracle of Apollo. There was a tradition also that Phocaea was not admitted into the Ionian federation until it took kings from the family of Codrus, last king of Athens.[3] Pyres and Parmenides may have claimed Phocis for their ancestral land. If they did not, many of their city folk did.

Phocaea must have had a difficult few years in the beginning. In

1. Herodotus, op. cit., I, 163–167.
2. Diogenes Laertius, op. cit., IX, 20.
3. Pausanias, *Description of Greece*, tr. W. H. S. Jones, et al. (London: Loeb Classical Library, 1918), VII, 3.

accordance with ancient practice many of the Greeks who emigrated to the east did not take women with them, but acquired them
by force from the indigenous population. Yet in many ways the first
Phocaeans were lucky. The site of their city had not been previously
inhabited, so nobody was dispossessed by their occupation. Phocaea
was the northernmost city of the Ionian federation and was later
considered an Ionian city in Aeolian territory, being north of the
Hermus river which was taken as the boundary between Aeolis to
the north and Ionia to the south. There was a story which explained
this, that the site of Phocaea and the country round it had been
given to the Phocaeans by the neighboring Aeolian city of Cyme.
This suggests that Phocaea was as closely allied to this city of the
Aeolian federation as to those of the Ionian. It was a generous gift, if
gift it was, because Phocaea had an excellent site and was excellently
placed. The Roman historian Livy describes it at about 200 BC.

> Phocaea is situated at the head of a bay. The width of the bay there is
> a mile and a quarter and from there a tongue runs out a mile into the
> sea, dividing the bay almost in the center as if with the stroke of a
> pen. Where it meets the narrow entrance it forms two very safe har-
> bours, to the north and to the south. The southern haven is called
> Naustathmus because it has room for an immense number of ships.[1]

This harbour was not the only natural feature of Phocaea which
enabled a rapid rise to power. Harbour and city were just north of
the mouth of the river Hermus, and closest to it of all the Greek colonies. The valley of the Hermus ran straight to Sardis, capital of
Lydia, in this period one of the richest cities of Asia Minor. Phocaea
was on a main route from Sardis to the west.

Greek colonial expansion in the east continued in the centuries
following the first emigrations and Phocaea and the other new eastern cities were its main agents. From about 750 BC Greeks began to
colonize the western Mediterranean. In this too Phocaea took a
leading part though it was not until the end of the westward expansion that Phocaea founded colonies of its own in the west. Herodotus tells us that the Phocaeans were the first Greeks to make long

---

1. Livy, *Livy, Rome and the Mediterranean*, tr. H. Bettenson (Middlesex: Penguin
Books, Ltd., 1976), XXXVII, 31.

sea voyages and so they had to be to venture west from Asia Minor rather than Greece. Phocaeans explored the coasts of Italy and Spain[1] and named Nice, Monaco, Antibes and Marseilles in the south of France.[2] All these were originally Greek names and the Phocaeans were the first Greeks to establish trading stations in this area. They went to sea not in broad beamed freighters but in fifty oared galleys, with a single rank of twenty five oars on each side. From the beginning these galleys probably had the ramming gear which they certainly had later. They were long low ships, made for fighting at sea but big enough to carry many kinds of goods, ships for breaking into new waters against maritime opposition. Not long before the Greeks founded their first colony in the west, the Phoenicians founded Carthage half way along the North African coast. The Phocaeans were Greek frontrunners in a race with the Phoenicians for the west. There were also the Etruscans already there. The Phocaeans of this period must have been daring men, deliberately going further than any Greek sailors before them. They must also have been better navigators and this may be connected with their claim to ancestral origins in Phocis. Delphi was in Phocis and Delphi was the geodetic center of the Greek world. From the center to the east and from the east to the west was the history of ancient Phocaea.

Let us consider Phocaea when Pyres, the father of Parmenides, was young. What conversations might he have heard on the quays of that magnificent harbour in, let us say, 560 BC? At this date Phocaea was at the height of its power, with trading stations and colonies at the four corners of the Mediterranean. In the north east was the Phocaean colony of Lampsacus, one of the cities on the Asian side of the Hellespont, a strait which runs from the Mediterranean towards the Black Sea.[3] The colony at Lampsacus enabled the Phocaeans to trade from their own base with the peoples on the coasts of the Propontis and the Black Sea. The land of the colony had been given, it was said, to a king of Phocaea by the king of the region in

1. *Livy, Rome and the Mediterranean,* op. cit., xxxiv, 9.
2. Boardman, op. cit., pp 226–227.
3. Ibid., p 249.

return for services he had received from the Phocaeans. This way of acquiring territory reminds us of the acquisition of the site of Phocaea itself which was also said to have been a gift, from the Aeolian city of Cyme. It is characteristic of the Phokaian expansion.

In the south was Naucratis in Egypt, on the Canopic mouth of the Nile, a large east Greek trading station which at about this time was granted a monopoly of the Mediterranean sea trade with Egypt by the Egyptian king Amasis. The Ionians, Aeolians and Dorians who traded there established a temple in common, the Hellenion, and Herodotus lists Phocaea among the members of the cult.[1] Between Lampsacus and Naucratis lay Phocaea itself and the value of these two bases north and south was in the territory they opened to trade, the worlds of the Black Sea and the Nile. Add to these the third world opened to the Phocaeans by a waterway, Lydia, Sardis and points east on the river Hermus just south of Phocaea, and we see how cosmopolitan the Phocaeans were and what range of cultures they habitually encountered. At this time the court of Lydia was especially open to the east Greeks, had acquired a Greek alphabet, sent offerings to Delphi and taught Greeks the use of money. King Croesus of fabulous wealth was on the Lydian throne, and the Greek coastal cities were at once his subjects and friends.

In the north west of the Mediterranean, on the south coast of what is now France, the Phocaeans had founded their first colony in the west, Massalia, modern Marseilles, about 600 BC Across the bay on the north east coast of Spain was a Phocaean trading station, and more again to the east of Massalia. Massalia had not been founded without a fight. Thucydides mentions the sea battle the Phocaeans fought with the Carthaginians at the time of its founding.[2] As usual they sited near the mouth of a waterway which gave access to a great territory in the interior, the river Rhône. There is a late but charming story that the daughter of Nanus, king of the region, fell in love with one of the two Phocaean leaders sent out from Phocaea to establish a base there. King Nanus gave the newly weds the site of

1. Herodotus, op. cit., II, 178.
2. Thucydides, *History of the Peloponnesian War,* I & II, tr. Charles Forster Smith (Cambridge, Harvard University Press, 1977), I, 13.

Massalia.[1] The volume of trade passing through Massalia in the early period must have been considerable, judged from the traces which have been found most of the way to Paris. There is evidence that Greek architects were working in this period near Munich.[2] The Massaliot Phocaeans may have tapped into the overland tin route from northern to southern Europe. They introduced the making of wine to France and found it a useful import in the first years of their dealings with the natives, who thought a jug of wine worth a slave.

In the center Phocaea, Sardis and Lydia; to the north Lampsacus, the Propontis and the Black Sea; to the south Naucratis and Nilotic Egypt; to the northern west Massalia and the French and Spanish trading stations. But the greatest achievement of Phocaea in 560 BC remains to be told. In the south west of the Mediterranean the Phocaeans had opened to trade the land beyond the pillars of Hercules, the straits of Gibraltar. About 640 BC a Samian ship on course for Egypt was blown by a Levanter so far as to pass through the straits of Gibraltar. It made landfall on the western coast of Spain in the kingdom of Tartessus. The Samians returned from there bearing one of the most valuable cargoes of any ship before them. But the Phocaeans and not the Samians exploited the discovery. Perhaps warships were needed against Phoenicians and Carthaginians. Phocaean sailors quickly established the most excellent relations with Tartessus and its king, the long lived Arganthonius. These people too may have been ignorant of wine or were happy to break a Phoenician monopoly in trade. Tartessus is hard to place but it is likely to have been near Cadiz, and as usual near the mouth of a waterway penetrating far into the interior, the river Guadalquivir. Tartessus was rich in silver and other metals and opened to the Phocaeans the new world of the Atlantic.

This then was the Phocaean trading empire as it was at the time of Pyres' youth. The city even had its own trademark, the seal or *phoca*. But Phocaean activity in the west remains anomalous. That the vanguard of Greek expansion in this direction should come from Asia

1. Kaeppel, op. cit., Justin, pp 116–117.
2. Boardman, op. cit., p 229.

Minor is strange enough, but that they should trade and not colonize like the other Greeks they helped is stranger. It is possible that the foundation of Massalia was no exception to this rule for Thucydides tells us that this colony was founded at the time of a sea battle with the Carthaginians. It may have been intended as a strong central defence for the trading stations in the Gulf of Lyon, and not as an emigrant colony. But at the time of which we are speaking this anomaly was removed. In about 565 BC at the command of the Delphic oracle Phocaea founded its second colony in the west, on the island of Cyrnus, modern Corsica. This colony was called Alalia, on the east coast and at the mouth of a central waterway. There is reason to suppose that Massalia played a part in the foundation of its sister colony. Alalia was a staging post for east Greek dealings with the north west and was also well placed for trade with two Etruscan cities on the Italian seaboard, Agylla and Gravisca. The colony at Alalia seems to have flourished without interference in its early years and its foundation is sometimes taken as the start of the Phocaean thalassocracy or naval empire. It also marks the limit of Phocaean expansion.

Shortly after the foundation of Alalia Phocaea itself was threatened by a powerful enemy from the east, the Persians. In 559 BC the Persian Cyrus took the throne of the Medes and Persians from his Mede grandfather Astyages. This change in the royal line of the Medes and Persians had far reaching consequences. Unlike the Medes the Persians were men of war. King Croesus of Lydia, on whose good will Phocaean prosperity and security depended, took the initiative by asking Apollo at Delphi whether he should attack Cyrus. The oracle's reply was ambiguous but Croesus took it as favorable and attacked. The battle at Pteria was drawn and Croesus withdrew to Sardis, to find Cyrus and his army at the gates soon afterwards. Cyrus took Sardis after a siege and captured Croesus. What was worse, Cyrus had asked the east Greeks to revolt from Croesus and fight with him but they had refused. Now they went to Cyrus in Sardis and asked him to receive them as subjects on the same terms Croesus had given them. Cyrus refused.

The Ionians and Aeolians on the mainland of Asia Minor then entered into an agreement for mutual defence against the Persians

and sent an embassy to Sparta to ask for help. Pythermus, a Phocaean, was chosen as spokesman. Herodotus describes Pythermus'
performance at Sparta, how he dressed in purple to attract a large
audience and spoke at length but was unsuccessful. It was a tactless
speech since the Spartans, in a group at least, valued simplicity in
manners and brevity in words more than all the other Greeks. But
no doubt Herodotus' parvenu Ionians wanted to impress the Spartans with their wealth and eloquence. In fact the embassy was not
completely unsuccessful. The Spartans sent a galley to Phocaea
from which a Spartan envoy went to Sardis and told Cyrus to leave
the Greek cities of the east alone. Cyrus asked who the Spartans
were and when he was told, treated their demand with contempt.
But Phocaea was taking other measures. King Arganthonius of Tartessus beyond the straits of Gibraltar offered the Phocaeans whatever
land they chose of his domains for a complete resettlement. When
the Phocaeans decided against this he gave them enough money to
build a splendid stone wall around their city as a defence against the
Persians.

Then Cyrus left Sardis for the east, taking his prisoner Croesus
with him. The east Greek cities were too insignificant to warrant his
personal attention who had Babylon and Egypt to conquer. He left a
Persian governor in Sardis and gave to the Lydian Pactyes the task of
gathering the wealth of Croesus and the other richest Lydians and
sending it on to him. Hardly had Cyrus left Lydia when Pactyes
raised a revolt among the Lydians and east Greeks against the Persians. Sardis was retaken and the Persian governor and garrison
killed. When Cyrus heard of this he ordered one of his generals to go
to Lydia, resubjugate it and sell into slavery all those who had helped
Pactyes. He was first to make a proclamation to this effect. This was
done, some Greek cities were taken and the citizens of Greek Priene
were sold into slavery. At this point Cyrus' general died and another
was appointed in his place, Harpagus, who immediately marched
against Phocaea.

The new walls of Phocaea were useless against Harpagus who laid
siege to the city and built an earth mound against the wall in order
to force an entrance. At the same time he announced to the Phocaeans that all he wanted was the pulling down of one of their wall's

towers and possession of a single house inside the city. But the Phocaeans believed, no doubt because of the proclamation and events in Priene, that his intention was to enslave them. Some or all of them must have taken part in the revolt of Pactyes against Cyrus' governor in Sardis. So they asked for a day to consider Harpagus' announcement and stipulated that he should withdraw his army from the walls for this period. Harpagus agreed but said he knew what they planned to do. Perhaps he did but there were now a number of courses open to the Phocaeans, not the least attractive of which must have been to surrender to Harpagus the Phocaeans who had taken a leading part in the revolt and do what else he asked. Instead they showed solidarity. As soon as Harpagus had withdrawn they launched their galleys, and after putting aboard their families, furniture and temple property, sailed away. When Harpagus came back after the time was up, it was to an empty city.

The Phocaean fleet sailed to the nearby island of Chios which was also Ionian. There, safe from Persian attack, they negotiated with the Chians for possession of the Oenussae Islands a little way from Chios. It is likely that the putting of this proposal to the Chians had been a condition of the Phocaeans' agreeing to leave Phocaea together, and it was a wise proposal on their part since the islands would provide them with a trading base in more or less the same position as Phocaea, and a base from which they could rapidly recover their city if the Persians withdrew. But the Chians refused to part with the islands for fear that a Phocaean base there would harm their own trade. The Phocaeans were forced to reconsider. Lampsacus on the Hellespont was as open as Phocaea to attack from Persia; Naucratis was a trading center only; Tartessus where Arganthonius had promised them land was now ruled by another king. They decided to go to their latest and nearest colony in the west, all of them, many hundreds of miles across the sea, to Alalia in Corsica. This must have been a hard decision to make but it was a harder one to impose. Before the great voyage west began, they returned to Phocaea and killed the garrison which Harpagus had left there. They then laid curses on any Phocaean who failed to go with the expedition and dropped a lump of iron into the sea, swearing never to return to Phocaea before the iron floated up again. To no avail. The voyage was just begun when over half the fleet turned back to Phocaea, breaking their oaths, defying the curses and risking Persian enslavement.

The others under the leadership of Creontiades kept their oaths and went on to reach Corsica safely.[1] Less than half a city's population is still a large body of people and they must have outnumbered the colonists already there. Unlike most Greek colonists the new arrivals were accompanied by old people, women and children whose presence made the colony more vulnerable and less efficient. Nonetheless all went well at the start; the new colonists built temples and settled down. It is here in the colony of Alalia at this time that we may at last place Pyres, the father of Parmenides, with some confidence. We can guess nothing else about him. Even by the earliest reckoning of Parmenides' birth date, his son was not to be born for another five years. Five years were all Alalia had left to it before its people were plunged into catastrophes far worse than those the new arrivals had already suffered and survived. Meanwhile the Phocaeans who had turned back to Phocaea became Persian subjects and lived on in relative peace.

In the years immediately following the arrival of the new emigrants Alalia, Herodotus tells us, plundered and pillaged its neighbors. This is uncharacteristic of Phocaean diplomacy as I have described it and we must wonder at the change. It is possible that the sudden increase in the population of Alalia required a more aggressive policy for its provision. But this is unlikely when we remember the wealth of Phocaea at the time of its evacuation, much of which must have come over with the new emigrants. Whatever else Alalia lacked, it was not the wherewithal to buy what it needed. It is more likely that control of the colony had passed to the leaders of the new emigrants who pursued the same vigorous and rash policy which they had in Phocaea. These were the men who had gone in with Pactyes when he revolted against the Persians. Fearing enslavement on the approach of Harpagus they had persuaded their fellow citizens to evacuate Phocaea and then made them take the oath to leave forever. They had already participated in the killings of

---

1. Strabo, op. cit., VI, I.I.

two Persian garrisons and one governor and were not overnice in their ways of winning agreement from their fellows. But Alalian piracy was soon to be checked once and for all. The Carthaginians and Etruscans made a temporary alliance against Alalia and sailed out to do battle with sixty ships each.

This time there was no escape. Recourse to the galleys with the women, children and old people against an enemy from the sea was impossible. They had sixty ships against one hundred and twenty but they were the best Greek sailors of their time and their ships were warships. They would have recalled, as they put out, their battle with the Carthaginian fleet at the founding of Massalia two generations before and prayed for a repetition of that success. Sixty galleys cutting across the sea of Sardinia in formation, on their way to fight the greatest sea battle in the Mediterranean of their century and to win immortal fame among the Greeks. Of the tactics and course of the battle we know nothing, only its outcome, the putting to rout of the Carthaginians and Etruscans. The Greeks won but only just and at great cost. Forty of their sixty ships were sunk and of the twenty still afloat the ramming gear was so badly damaged that they were incapable of further combat. These crews returned to Alalia, forced to abandon their comrades from the sunken ships in the water. We may imagine the despair in Alalia at their news and as they sought their families to take them aboard and away before the enemy arrived. Hector seeking Andromache in the streets of Troy, [1] Aeneas with his family in its burning.[2] About this time is the earliest date given for Parmenides' birth.

The twenty ships, the crews and their families managed to escape from Alalia and safely crossed the sea to reach Rhegium, modern Reggio, a Greek colony on the toe of Italy. Here they were allowed to make a temporary base. As for those who were left in Alalia we may hope they survived. There is some evidence that they emigrated to Sardinia. In any case the colony at Alalia was abandoned. But there was still one more disaster for these unfortunate Greeks to bear. The Carthaginians and Etruscans captured many of the Greek sailors

1. *Iliad*, vi.
2. *Aeneid*, ii.

who had been left in the water after their ships were sunk in the battle. The captors then drew lots to decide who should own these prisoners and among the Etruscan winners by far the greatest number was won by the city of Agylla. The Agyllans then took their prisoners ashore and stoned them to death. Herodotus gives no explanation of the action nor is it easy to plumb what effect this latest news would have had upon the Alalian survivors. But this atrocity was to mark the limit of their suffering as the founding of the colony at Alalia had marked the limit of Phocaean success a generation before. For there was an extraordinary sequel:

> The result of this outrage was that when any living thing - sheep, ox, or man—subsequently passed the place where the Phocaeans had lain, its body became twisted and crippled by a paralytic stroke. Wishing to expiate the crime of the murder, the men of Agylla sent to Delphi, and were told by the priestess to begin the custom, which they still observe today, of honouring the dead men with a grand funeral ceremony and the holding of athletic and equestrian contests.[1]

This is what Herodotus says and we may believe as much of it as we choose. There can be no doubt that the Agyllans did establish the games and the funerary rites since Herodotus says that they were continued to his own time nearly a century later. There can be no doubt either that they were established in honour of the sailors who fought in the sea battle and were murdered, nor that they were authorized by Apollo at Delphi through his oracular priestess. We may be sure then that many of the generation of Parmenides' elders had been canonized by the god himself and were honoured by festivals throughout Parmenides' lifetime. As for the dramatic sickness induced in passers by, it does not sound much like a contagion from rotting corpses as some suggest. To the Greeks it would have seemed proper that these dead heroes should be the concern of gods who had protected the corpses of Hector from Achilles[2] and of Polyneices from Creon.[3] We can know no more of the necessity which

1. Herodotus, op. cit., i, 165.
2. *Iliad,* xxii–xxiv.
3. Sophocles, *Oedipus the King, Oedipus at Colonus, Antigone,* vol.1, 'Antigone', tr. F. Storr (London: Loeb Classical Library, 1981).

drove the Agyllans to defer to the authority of Delphi in this matter than Herodotus tells us. Delphi's response was to appease the dead and conciliate the living.

For some time the surviving Alalians remained at Rhegium. A generation before, the god had directed them to found their colony in Cyrnus, Corsica. That colony they had been forced to abandon. Were they, with so many men dead, to attempt to return there? Or should they ask to become citizens of Rhegium? Or consult the oracle again? Perhaps the god had already declared their dead kinsfolk heroes. To this problem too a solution was found which in the short and long term proved acceptable to everyone. A man from Poseidonia (Roman Paestum), a Greek city on the west coast of southern Italy, suggested to the survivors that when Apollo had told them to found their colony, it was not the island but the hero Cyrnus he had meant, and the Alalians had only to establish a cult of this hero to comply with the god's requirements. This suggestion the Alalians adopted. It enabled them to fulfil the god's command to the letter and at the same time settle where they liked. So they sent out settlers to found a new colony on the west coast of southern Italy, twenty five miles south of Poseidonia. This colony was Elea, named perhaps after Elaea, a Greek city near Phocaea.

The choice of site indicates more help from the people of Poseidonia, who seem to have done for the refugees from Alalia something of what the people of Cyme had done for their ancestors at the founding of Phocaea. The good luck of the Phocaeans was returning. But they were very cautious. The site was the top of a high and very steep hill overlooking the sea.[1] Here they laid out their new city and strongly fortified it. Its position guaranteed them against mounds. Below was a bay and on either side of the hill a river flowed into the sea, the two mouths of which were harbours. To the north was the great city of Poseidonia. Beyond that was Italian Cumae, the first Greek city founded in the west. Beyond Cumae were Etruria and the festivals of Agylla. To the south east were the Greek cities Metapontum, Taranto and luxurious Sybaris. More to the south were Croton where Pythagoras lived, and their friends in

1. Boardman, op. cit., p 200.

Rhegium. Elea was a small, well protected and wealthy foundation. Its people, Herodotus tells us, were the only east Greeks apart from the Teans who preferred voluntary exile to the prospect of Persian enslavement. Having won their autonomy the Eleans took no further part in great affairs and from this time their name hardly appears in the history of states. But in the history of philosophy Elea was to give its name to several of the most famous philosophers in the ancient world, the Eleatic school. Of this school Parmenides was and is accounted the leading representative and we shall see how the history of his people informs his thought. For it is to be expected that the immediate past and immediate future of these people, at the time of their city's founding, were closely related.

The fullest account of Parmenides' own life is given by Diogenes Laertius in his *Lives of the Philosophers*, written in the second or third century A D. If we subtract what Diogenes tells us there of Parmenides' opinions we are left with the following:

> Parmenides of Elea, son of Pyres, was a pupil of Xenophanes. But though he was instructed by Xenophanes he was no disciple of his. According to Sotion he also associated with Ameinias the Pythagorean who was the son of Diochaitas and a worthy gentleman though poor. This Ameinias was the man whose disciple Parmenides became and when Ameinias died Parmenides built him a hero's shrine, being himself of rich and famous birth. And it was by Ameinias, not by Xenophanes, that Parmenides was converted to the contemplative life. . . . Parmenides flourished in the sixty ninth Olympiad. . . . He is said to have served his city as a legislator.[1]

It is not much but it is enough. The place, Elea, and the time of birth, about 540 BC, would tell any ancient reader who knew the first book of Herodotus' *Histories* who Parmenides was among the Greeks. We should rather be grateful for the brevity of the biography when we compare it with the numerous ancient hagiographies of Pythagoras or with the absurd life of Heraclitus, which is cobbled together out of ironic misinterpretations of his maxims. Parmenides' doctrines are no less apparently absurd than those of Heraclitus and it is surprising that he did not attract similarly scurrilous

---

1. Diogenes Laertius, op. cit., IX, 21–3.

anecdotes. But he avoided both the encomium and the satire, perhaps because he had a clear historical identity already provided by the pages of Herodotus.

If we subtract further from Diogenes' account what he or his predecessors could have deduced from the pages of Herodotus, knowing only that Parmenides was Elean and his time of birth, we are left with even less. The rich and famous family of Parmenides is rich merely because Phocaean and famous merely because Elean. He made laws for Elea because he was an important thinker living not long after the foundation of his city. Diogenes may have deduced that he was taught by Xenophanes because he thought that Xenophanes had written a poem on the settlement of Elea,[1] was of the previous generation and had described himself as a traveller.[2] As for Ameinias, Parmenides' other teacher, Diogenes tells us that Sotion said that Parmenides built him a hero's shrine. We know no more of this Ameinias than Diogenes tells us here. Let us suppose that there was a shrine with an inscription in later times. That, the pages of Herodotus and some knowledge of Xenophanes are sufficient to explain almost everything Diogenes tells us about Parmenides' life. But that, of course, does not mean that Diogenes is wrong. The inferences are reasonable.

Xenophanes' poem on Elea would have told the story of the earlier part of this chapter, and it is possible that Herodotus derived information from it. Xenophanes, Diogenes tells us, was a native of Colophon, another of the Ionian cities on the seaboard of Asia Minor, south of Phocaea. In one of the surviving fragments of his work Xenophanes tells how he left home at the age of twenty five and spent the next sixty seven years travelling over the Greek world. He cannot then have settled long in one place. In another fragment he describes the men of Colophon:

> After learning useless luxuries from the Lydians, so long as they were free from hateful tyranny, they used to go to the assembly in purple robes, not less than a thousand of them in all, haughty men adorned

1. Diogenes Laertius, op. cit., IX, 20.
2. *Ancilla to the Presocratic Philosophers,* op. cit., Xenophanes, frag. VIII.

with well dressed hair, steeped in the scent of carefully prepared unguents.[1]

The purple robes have the same effect on Xenophanes as the purple robe of Pythermus the Phocaean on the Spartans. These verses would have been popular in Elea if the Eleans regarded their former compatriots in Phocaea, now under Persian control, with any lack of respect. Xenophanes' experience was like their own. His roving and his dislike for the oppressive life of the east matched theirs, while their adventures were a proper theme for his poetry. He was by no means ascetic despite his criticisms of the Colophonians:

> One should say things like this by the fire in winter, lying on a soft couch, well fed, drinking sweet wine and nibbling chickpeas: 'Who are you among men, and where from? How old are you, my good friend? What age were you when the Mede came?'[2]

It was a great pleasure for recent emigrants from the east to discuss the Medes and Persians, more pleasant even than the wine and chickpeas.

Xenophanes was a generation older than Parmenides according to Diogenes and we have seen that he was still writing verse in his nineties. This long span of active life makes it difficult to determine from his verses what his influence was on the younger man, since some and perhaps many of them were written after Parmenides had come to maturity. It is possible that Xenophanes was as much influenced by Parmenides in his later works as Parmenides by him, and this may have been why Diogenes described Parmenides as a pupil of Xenophanes but not his disciple. There are some striking similarities between the poems of the two men. Both adopted the epic hexameter as a vehicle for their philosophy, a metre which both managed with a gnomic precision.

Compare for example a single hexameter by Xenophanes on God:

> He sees as a whole, hears as a whole, thinks as a whole.[3]

with an hexameter of Parmenides on being:

---

1. *Ancilla to the Presocratic Philosophers,* op. cit., Xenophanes, frag. III.
2. Ibid., frag. XXII.
3. Ibid., frag. XXIV.

Nor is anything less, but everything is full of being.[1]

These lines illustrate another similarity between the poets, their approach to divine matters. Xenophanes is best known for his attacks on the anthropomorphism of Greek theology, its tendency to conceive of gods with all the faults of people. But Xenophanes is not an atheist. On the contrary, he argues for a purer and loftier understanding of God and adopts the position of a greater piety. Parmenides takes the same position. Though his poem begins in the anthropomorphic manner with his coming to a goddess, what she tells him about is being, which seems to have little in common with the Homeric gods of Olympus.

Xenophanes and Parmenides are also alike in being associated with the school of Pythagoras. According to Diogenes, Xenophanes was the same age as Pythagoras who came over to Croton in southern Italy from Samos off the coast of Asia Minor. There is a fragment of Xenophanes, possibly ironic, in which Pythagoras is described as having told a man not to beat a dog because from its cry he had recognized the soul of a friend.[2] Parmenides was a disciple of the Pythagorean Ameinias for whom he built a hero's shrine and of whom we know nothing more than Diogenes tells us, not even his city. But if he taught Parmenides he was probably taught by Pythagoras at Croton. His poverty and good birth together could be explained by his having given his wealth to the school there. The building of the shrine to him is of a piece with the rapid development of new cults by and around the Eleans. But the fact that Parmenides was a disciple of the Pythagorean Ameinias tells us as little about the origins of his thought as that he was a pupil of Xenophanes. If Parmenides was a Pythagorean at this time, this tells us more about the Pythagoreans than it does about him. For we have a number of verses by Parmenides but nothing by Pythagoras nor by the first Pythagoreans.

Parmenides' most famous disciple was also Elean, Zeno of the paradoxes. Diogenes tells us that Zeno was the son of Teleutagoras by birth but of Parmenides by adoption, that he was thoroughly trained by Parmenides and was his boyfriend.[3] Zeno was one of the

1. Parmenides, op. cit., frag. VIII.
2. Xenophanes, op. cit., frag. VII.

most remarkable thinkers of his time, but his was a mind which in its expression at least was unlike that of his master. Zeno was a dialectician while Parmenides was both dialectician and poet. Nonetheless Zeno's work was supportive of his master's as it defeated his master's opponents by reducing their arguments to self contradiction and absurdity. This was the purpose of his paradoxes of which the best known is the one about Achilles and the tortoise. Parmenides had said that being was immobile and Zeno showed that motion was impossible. But if Zeno was not a poet like Parmenides and Xenophanes, he had learnt their art of gnomic precision:

Anything moved is moved neither where it is nor where it is not.[1]

The dialectics of the Eleatic school were also developed by Melissus of Samos.

The prose work of Zeno and Melissus has profoundly determined the modern view of the Eleatic school. The poetry of Xenophanes and Parmenides is now regarded as an ornament obscuring their contributions to philosophy. The poetic tradition of the Eleatic school was carried on by Empedocles of Acragas in Sicily, the last poet in our era regarded as having been an original natural philosopher. Empedocles' hexameters are full of echoes from the verses of Parmenides and he was also a Pythagorean. He was probably an admirer of the earlier poets rather than a disciple who studied at their feet. The late sixth and early fifth centuries BC, the time of Xenophanes and Parmenides, produced a large body of gnomic verses by various hands, some of which were collected into a single work and attributed to the poet Theognis. We know nothing of Theognis for certain except that one of the poems in the collection explains how poems addressed to Cyrnus, as this one is, will be attributed by everyone to Theognis the Megarian.[2] There has been debate about which Megara is meant, the Sicilian or the mainland Greek, but what interests us is the name Cyrnus. It is the name of the hero to whom the Eleans established a cult on the advice of the man

---

3. Diogenes Laertius, op. cit., IX, 25.

1. *Ancilla to the Presocratic Philosophers*, op. cit., Zeno, frag. IV.

2. *The Penguin Book of Greek Verse*, ed. C.A.Trypanis (Middlesex: Penguin Books, Ltd., 1972), Theognis, 19–38.

from Poseidonia. The Cyrnus of the poems is addressed as though he were the poet's boyfriend, but while there are many poems to Cyrnus we do not learn much from them about Cyrnus himself.

Who was the Cyrnus whom the Eleans worshipped? The hero after whom the Greeks named the island of Corsica. So much we may deduce from what Herodotus tells us. Diodorus, a Sicilian Greek writing at the time of Julius Caesar, tells of a Cyrnus who was sent by the legendary king Inachus of Argos to find his daughter Io. Cyrnus went all over the inhabited world in a vain search for Io and then settled in Asia Minor, since Inachus had told him not to return to Argos without her.[1] This may be the Cyrnus after whom the island of Corsica was named, an appropriate hero for Phocaeans. The Hellenistic poet Callimachus ranked the island of Cyrnus second only after Delos of all islands in his poem on the birth of Apollo on Delos.[2]

There is one other possibility. The western Greeks like many other emigrants in new surroundings tended to use the place names of the old country. Thus on the west coast of Italy we find Cumae as in Asia Minor there was Cyme as in Greece there was Cumae. The debate over the Megara of Theognis is another case. Close to Phocaea in Aeolian territory was the city of Grynea, famous for its temple and grove sacred to Apollo Gryneios. There were other temples to Apollo Gryneios on and off the coast of Asia Minor. When Trojan Aeneas justified his desertion of Dido, queen of Carthage, he argued according to Virgil that Apollo Grunius had allotted him the task of founding a city in Italy.[3] Gryneios and Grunius are adjectives formed from the name Grynus, the hero said to have founded the Aeolian city of Grynea.[4] It is just possible that the Cyrnus of the Eleans is this same Apolline Grynus they had known in Phocaea. There was also Apollo Carneius, named after the hero Carnus whom the Spartans worshipped as a founding deity and to whom there was a sacred grove dedicated.

1. Diodorus Siculus, op. cit., v, 60.

2. *Hymns and Epigrams*, op. cit., Callimachus, IV, 19.

3. Virgil, *Aeneid*, tr. H. Fairclough, Rev. G.P. Gould (London: Loeb Classical Library, 1961), IV, 345.

4. Herodotus, I, 149. See also Servius on Virgil *Eclogues* VI, 72.

THE FRAGMENTS
OF PARMENIDES

I

The horses which carry me as far as my heart may reach were escorting me when they brought me and came to the resounding road of a demon goddess, which carries the man who knows through all cities. There was I carried, for there the wise fillies carried me as they strained at the car, and maidens were leading the way. The axle in the boxes let out the shriek of a pipe, burning, for it was driven round by the two whirling circles at both ends whenever the maidens, daughters of the sun, hastened their escorting after leaving the houses of Night for the light and pushing back from their heads their veils with their hands. There are the gates of the paths of Night and Day, and they have a lintel around them and a stone threshold. Ethereal, they are filled by great doors and of these much punishing Justice holds the alternate keys. Her the maidens appeased with soft words and carefully persuaded her to push out for them the rounded bar from the gates. Then they opened and made a yawning chasm of the doors, revolving the brazen doorposts, studded with bolts and nails, one after the other in their sockets. Straight through them the maidens took horses and car down the road. And a goddess gladly received me and took my right hand in her hand and said this to me:

'Young man, companion of immortal charioteers, the horses which carry you as far as our house, Welcome! Because it was no evil fate which predestined you to take this road, for it is far indeed from the path of people, but Right and Justice. You must discover all, both the unshaken heart of well rounded truth, and also the opinions of mortals in which there is no true trust. Nonetheless you must learn of these as well, how the things which seem had certainly to be, all of them passing through everything.'

II

Come, I shall speak, and when you have heard my speech, study it. These are the only roads of enquiry of which we may conceive. The one road is that it is and that it cannot not be; this is the path of persuasion for it follows on truth. The other road is that it is not and that it is necessary that it not be; I tell you that this path is absolutely unknowable. For never could you know what is not, because that is impossible, nor could you say it.

III

For to think and to be are the same thing.

IV

But look how things which are absent are firmly present to the mind nonetheless. For you will not cut off what is from holding onto what is, which is neither all scattered everywhere in order nor put together.

V

It is all one to me where I begin, for there shall I come back again.

VI

You must say and think that what is is. For being is and nothing is not. These things I bid you ponder. For this is the first road which I forbid you to take, and next there is the one which mortals wander down, knowing nothing, two-headed. For an impossibility directs the wandering mind in their hearts. They are carried along deaf, blind and aghast, uncritical crowds by whom being and not being are thought to be the same and not the same. And of them all the path is backward turning.

VII

For never shall this prevail, that things which are not are. Hold back your thought from this road, and do not let practised habit force you down it, to roam with unseeing eye and echoing ear and tongue. But judge by reason the much tried proof I speak.

VIII

It only remains to speak of the road that it is. On this road there are many signs indeed, how being is ungenerated and indestructible, whole, unique, unshaken and perfect. Nor was it ever, nor will it be, since it is altogether now, one and continuous. For what birth will you seek for it, from what has it developed to what? I will not let you say nor think that it comes from not being. For there is no saying nor thinking that it is not. What need, early or late, could have made it begin from nothing and grow? Being must either be altogether or not. Strong conviction will never allow anything to come from what is not besides itself. This is why Justice has not released being from its chains, nor allowed it to be generated or destroyed, but holds it fast.

The verdict in these matters rests on this: it is or it is not. And so our judgment must be to quit the one road as inconceivable and nameless, for it is no true road, and to take the other as real and true. How could being be at another time? How could it come to be? If it came to be, it is not, nor is it if it is ever going to be. So generation is extinguished and destruction unknowable. Nor is being divisible since it is all alike. Nor is there more of it here than there which would prevent it from holding together. Nor is there anything less, but everything is full of being. In this way it all holds together, what is is close to what is.

Immobile in the limits of great bonds it is without beginning or end, since generation and destruction have wandered far off indeed and firm belief has banished them. Remaining the same in the same place it rests by itself and so remains fixed. For strong Necessity holds it in the bonds of limit which constrain it all round, because it is right that being not be incomplete. For being lacks nothing. If it lacked anything, it would not be all.

But thinking and what it thinks are the same. For you will not find thinking without being, in which it is expressed. For neither is there nor will there be anything besides being, since Fate chained it to be whole and immobile. To it belong all the names which mortals established in the belief they are true, birth and destruction, being and not being, changing of place and of color.

Since it has a furthest limit it is perfect all over, like the mass of a well rounded sphere stretching equally in all directions from the center. For there cannot he anything which is greater or more powerful in one part of it rather than another. Nor is there anything which is not, which would prevent it from reaching its like. Nor is being more here and less there, since it is all inviolate. For whatever is equal in all directions is no less set within limits.

Here I stop speaking and thinking about truth in a way to be trusted. Learn after this the opinions of mortals by attending to the deceitful order of my words. Mortals decided to name two forms of which there was no need that there be one, and in this they were in error. They distinguished between these forms and gave to each signs which were separate from each other. To the one the ethereal fire of flame, gentle and very nimble, everywhere the same as itself and not the same as the other. The other also was all by itself and opposite to the first, unknowing night, a dense and heavy body. I tell you the whole likely order so that mortal thought may never outstrip you.

IX

But when all things are named light and night, and these two according to their powers are attributed to different things, then all is at once full of light and obscure night, of both equally, since there is nothing not shared by both.

X

You will know the nature of the ether and all the signs in the ether and the invisible works of the pure torch of the bright sun and whence they were born. You will discover the nature and wandering works of the round faced moon, and you will know too the surrounding heaven, from where it grew and how Necessity brought it and chained it to hold the limit of the stars.

XI

...how earth and sun and moon, the general ether and the heavenly milk, the outermost Olympus and the hot strength of the stars rushed to be born.

XII

The narrower crowns are filled with unmixed fire, those next to them with night, and a portion of flame is sent between. In the middle of these is the demon goddess who governs everything. For she rules everywhere over hateful birth and mating, sending the female to mate with the male and male to female back again.

XIII

... devised Love the very first of all the gods.

XIV

... shining at night, wandering round the earth, borrowed light.

XV

... always gazing upon the rays of the sun.

XVI

For as each time there is a mingling of the far wandering parts, so has mind come to people. For it is the same thing which mind thinks, the nature of the parts, in each and every person. For what fills is thought.

XVII

... young men on the right, young women on the left.

XVIII

When a woman and man mix the seeds of Venus at the same time, the power from the different blood works in the veins and fashions well-formed bodies if it preserves due proportion. For if the powers fight in the mingled seed and do not make one power in the mingled body, they will grievously trouble the growing sex with twin seed.

XIX

So, according to opinion, did these things grow and are now; and so, once nurtured, shall they afterwards die. For each of them people established a sign.[1]

---

1. Text from Leonardo Taran, *Parmenides* (Princeton: Princeton Univ. Press, 1965) except for frag. VIII 12 from Simplicius.

2

JUSTICE

NINETEEN FRAGMENTS are all that remain to us of Parmenides'
work and one of them is a Latin translation (XVIII). No text of the
complete poem has come down to us from antiquity. The same is
true of the poems of Xenophanes and Empedocles. Instead we have
quotations from their poems in the works of other ancient writers,
and the task of scholars has been to collect these quotations, collate
them and put them into the order, despite gaps, from which they
appear to have come. Nor is it easy to say, once this has been done,
how much of the original remains to us. In the case of Parmenides'
poem it is generally supposed that we have the whole of the introduction, most of the second part concerning being, but only a small
fraction of the last part. We can however be fairly sure of the order
of Parmenides' fragments because they come from a single poem
and one which has a clear and conventional structure.

The poem opens with the description of a chariot ride which
takes the poet to the palace of a goddess who tells him the rest of the
poem. This introduction is a more elaborate form of Homer's invocations of the Muse at the beginnings of the *Iliad* and *Odyssey*.[1] It is
strictly the Muse who sings the poems, the poet is there as himself
only to introduce her. More like Parmenides' introduction is the
longer beginning to Hesiod's *Birth of the Gods* where Hesiod sings
of how the Muses came down to him while he was shepherding his
lambs near Mount Helicon and inspired him to relate the generation of the gods.[2] Parmenides is a poet more like Hesiod than
Homer in this too, that his poetry is didactic rather than narrative,

---

1. *Iliad*, I, 1 and *Odyssey*, I, 1.
2. *Hesiod, The Homeric Hymns and Homerica*, op. cit., 'Theogony', 1–35.

its message not suitable for story. But the message is very clearly set out by the goddess in her palace who indicates in several places just where she has reached in her exposition and what she is to speak of next. And this despite her saying that it makes no difference to her where she begins since she will be coming back to there (V). It may make no difference to her but her clear directions to her argument are of the greatest help to us who have no better means of ordering the fragments we have of it.

The poem opens with a description of a chariot ride which brings a rider to a goddess who is to teach him. Given the need for an opening we might have expected a sea journey as more apt to the experience of Parmenides' people. But there are echoes of that experience throughout the poem. The rider travels a road which takes the man who knows through every city. Of the Greeks of their time the Phocaeans were the most widely travelled, and the Eleans still more so who had for some time been without a city of their own. The chariot ride is potentially frightening to the rider. The goddess in the palace tells him to take heart, it is a good and not evil destiny which has brought him to her. We remember the long journey from Phocaea to Alalia of the evacuees and the subsequent escape from Alalia in twenty damaged ships and the enemy fleets abroad. The words of the goddess guarantee safe conduct and Eleans would know the force of them better than most.

There is, I think, an allusion to the Phocaeans who broke their oaths to sail to Alalia and returned to Phocaea in the denunciation of people who know nothing, for whom to be and not to be are the same and not the same (VI). These Phocaeans swore oaths by the gods and then went back on them, which is as much as to say that they first distinguished between the real and the unreal and then confounded the two. The path of them all was backward turning literally and metaphorically. On this interpretation these lines are similar to Xenophanes' attack on the luxurious habits of his former compatriots, the Colophonians. But that was Xenophanes himself speaking, so far as we know. Here it is a goddess. If these words apply to the forsworn Phocaeans they are condemned by the goddess for turning back as the rider is encouraged in his having taken a road far from the path of people. This connects the circumstances which led to the founding of Elea with its rapid rise to philosophical prominence. The Eleans were granted a special revelation as a reward for their persistence. Because they had held fast to the real and kept straight on, they were permitted to see the real more clearly, a privilege accorded by Justice who brings and admits the rider to the palace of the goddess who tells him about it. Perhaps, finally, there is a connection between Parmenides' chariot ride and the horse races established by the Agyllans in honour of the dead sailors.

Whatever the historical and philosophical implications of the poem, it opens with a fine description of a chariot ride from the point of view of a rider. Everywhere there is the noise and motion of things performing at their peak. The axlepost shrieks in the boxes; the horses are free of the rider's control; the great doorposts of the palace, studded with bolts and pins, revolve in their sockets with a mechanical perfection. Alarm and exhilaration, beauty which is the terror we can still just bear. This is the authentic epic world we recognize from Homer. Everything is well made or well done, the very best of its kind, glittering and resonant at the limit of its potential. Who has ever gone as far or as fast as his heart could desire? There is a road, certainly, and it resounds but it is the road of a demon goddess; there are maidens but they are the daughters of the sun; there are gates but they are the gates of Night and Day; and there is a palace but it is the palace of a goddess. For Homer too there is no difference in kind between the worlds of gods and men, and so the world of men is made ideal. There is nothing here which cannot serve as a support for the divine in the perfection of its earthliness. And that being so, we must take the chariot ride seriously. However much we derive from it symbolically or analogically, it is still a chariot ride to a palace. In the same way, whatever a ritual means, it is the ritual that matters. It generates the meanings, not the other way round. The chariot ride is a powerful poetic symbol because it could be an actual symbol. There is nothing in the opening lines which could not have been ritually enacted. There are several poems of this period and in this metre which describe rituals. It is likely that this is one of them, and that the chariot ride was an actual and repeated event. It is a kind of triumph.

An unusual feature of these opening lines is the indefiniteness of the people, animals and objects they describe in their relations to each other. Consider the horses. They are female, fillies or mares, and they rush along the road which they know, straining at the car. Then suddenly there are maidens leading the way. Then we hear of neither horses nor maidens through the lines concerning the axle. Two teams, both of females and both doing the same work. It is hard to keep the maidens distinct from the horses. The horses are wise in their knowledge of the road, the maidens wisely persuade Justice to unbar the palace doors. The maidens talk but then so do the immortal horses of Achilles in the *Iliad*.[1] These similarities would not be remarkable by themselves but they are compounded by others between the parts of the chariot and the parts of the palace. The burning chariot axle is turned in the boxes by the two wheels, the bronze doorposts are turned in their sockets by the opening of the two doors. The chariot has one axle and the palace two doorposts, but it is the similarity between axle and doorposts which the poet seems to stress. For example he describes in full how the two wheels of the chariot drive the axle round, which elaborates the parallel with the turning of the doorposts by the opening of the two doors. In the Greek this is more noticeable still since *axon* appears in both descriptions, first meaning axle and then doorpost; and so too does *syrinx*, first meaning pipe and then socket. In this way the horses are transformed into the maidens, the chariot wheels into the palace doors, the axle into the doorposts, the boxes into the sockets, the burning into the bronze. Chariot and rider enter the palace with horses and maidens, bringing everything together at the palace of Night. But everything has been brought together in another way too, by the ambiguous and echoing description.

The indefiniteness of these lines is unusual and we ask whether the poet intended it. The goddess in the palace gives some reason for supposing he did when she tells the rider at the end of the first fragment to learn how the things that seem are all of them passing through everything. On our interpretation of these lines so far, we can see how they achieve an interpenetration of things in two different

1. *Iliad,* xix, 408.

200 SCIENCE AND RELIGION IN ARCHAIC GREECE

ways. There is the straightforwardly physical collection of all the
people, animals and objects at or in the palace of Night. The passage
of chariot, maidens and horses through the gates and into the palace
is the prime instance of this physical penetration. Others are the
passings of axle, keys, door bar and posts through their various
apertures. Then there is the merging of horses with maidens and
chariot with palace through their similarities to each other. These
ways combined would still not be enough to effect a complete passing
of all the things which appear in these lines through all. Not everything here becomes everything else; perhaps in the end nothing quite
does. But if it is the purpose of these lines to achieve the interpenetration of things as far as they may, then the indefiniteness I have
pointed to has its place. It is no more unusual than the poet's purpose
which the goddess in the palace describes and which is effected in
other ways by story and metaphor.

Who is the goddess who receives the rider and tells him the rest
of the poem? Her name is not given because there is no need to give
it. The inference is clear. The first words of this goddess address the
rider as the companion of the horses which have brought him to her
palace. There is only one mention of a palace or house before this,
or indeed anywhere else in the poem as we have it, and that is of the
palace of Night which the maidens leave to escort the rider to the
goddess. So the goddess in the palace must be Night. Night is a very
ancient goddess whose house is also described in Hesiod's Birth of
the Gods.[1] It is fitting that the speaker of the rest of the poem be
Night because Night was an oracular goddess who gave words of
wisdom to the people who went to her home for them. The chariot
ride, we may say, is the process of consultation. The only problem is
where exactly Night's house was, since it was a matter of debate
among the ancients. Hesiod seems to place it in Hades, near the
Styx. Plutarch, writing in the first century AD, tells us in a myth that
Orpheus had said that Night and Apollo shared the oracle at Delphi.[2] But according to the myth Night and Apollo have nothing in
common. Instead the oracle of Night has no single location but

1. *Hesiod, The Homeric Hymns and Homerica*, op. cit., 'Theogony', 744–757.
2. Plutarch, *Moralia*, 'On the Divine Vengeance', 566c.

wanders about. Plutarch was himself a priest at Delphi and this gives weight to his view of the issue.

But it was an issue, and the myth's resolution is drastic since it has to attribute a faulty memory to the great Orpheus. Euripides tells a different story.[1] According to him, when Apollo came to take possession of the Delphic oracle from the earth goddess Gaea, she let him and then played a trick on him. She sent up prophetic dreams at night everywhere on earth and so nobody needed to consult the Delphic oracle anyway. Apollo went to Zeus who told Gaea to stop it, and all the nocturnal oracles went back where they came from. The Pythian prophetess at Delphi, the Sibyls of Asia and Italy were living embodiments of Parmenides' goddess Night. They too spoke their oracles in hexameters. According to one of Diogenes' sources Pythagoras was taught by the prophetess Themistoclea at Delphi.[2] Parmenides was a disciple of Ameinias who was a disciple of Pythagoras.

Truth is told by a maddened woman in an underground cave. Night was at the very center of early Greek thought. Not until Plato wrote *The Republic* in the fourth century BC did light become the primary image of truth which it has remained ever since. Before Plato it was mostly the other way. Homer was reputed blind, his minstrel Demodocus in the *Odyssey* was blind.[3] His wise man Teiresias was also blind[4] and according to one tradition had been a woman as well as a man. Epimenides slept for decades in a cave,[5] Heraclitus was known as the dark because of the obscurity of his doctrines. There was Sophocles' Oedipus who thought he could rule Thebes by his own wits and the courage of his manhood, but who discovered in the end what the Delphic prophetess and Teiresias had known all along. He blinded himself.[6] These examples are

---

1. Euripides, *Trajan Women, Iphigenia among the Taurians, Ion, Vol. IV*, tr. D. Kovacs (London: Loeb Classical Library, 1999), 'Iphigenia among the Taurians', 1234–1283.
2. Diogenes Laertius, op. cit., VIII, 8.
3. *Odyssey*, VIII, 64.
4. Ibid., X, 493.
5. Diogenes Laertius, op. cit., I, 109.
6. Sophocles, op. cit., 'Oedipus the King'.

of truth as dark not light, they emphasize the inadequacy of light as a way of knowing. Plutarch suggests that Orpheus was wrong about the oracle shared by Night and Apollo at Delphi. But Sophocles had felt and evoked just such a tension between the dark and light of Delphi a long time before Plutarch. The emphasis on darkness in the earlier period was an indication of the sacred nature of knowledge, its inaccessibility to merely human understanding, its requirement that we approach on its terms, not our own. It is a marvellous thing that the first and arguably clearest account of being which we have is the doctrine of a dark lady. She is the one at the heart of well rounded truth, the rider is there only while with her. The speech of Night is often accounted the first philosophical reasoning in the west which has survived.

Now Justice, the lady of punishments who keeps the changing keys to the palace of Night. Justice is a goddess registered in Hesiod's Birth of the Gods[1] and so too is Themis[2] who is named with Justice by the goddess in the palace. It is the work of these two goddesses, Right and Justice, and not evil destiny to bring the rider to the palace of Night. Justice controls both the gates at which he arrives and his arriving at them. She may also be the demon goddess of the very first lines whose road the horses take to bring the man who knows across every city. Then Justice unlocks the gates with one of her keys and lets him in.

The demon goddess whose road the horses take is mentioned again in what remains to us of the last part of Parmenides' poem. Here she is described as being in the midst of crowns of unmixed fire and night (XII). These crowns have been thought to be bands of alternating fire and darkness around the earth and extending outwards, the fiery bands at different heights the courses of stars and planets. The fiery bands on this account are narrower because separated from each other by wider bands of darkness, and the demon goddess in the midst of all the bands is either half way between the earth and the outermost limits of the stars or at the center of the whole system. Aetius, a late commentator on Parmenides' poem,

1. *Hesiod, The Homeric Hymns and Homerica*, op. cit., 'Theogony', 748–750.
2. Hesiod, ibid., 902.

ascribes to Parmenides a theory of mixed crowns also, between those of fire and night and composed of both. In the very midst of these, he says, is the demon goddess to be found.[1]

There is another way of reading Parmenides' lines which explains Aetius' comment. The crowns of unmixed fire and night may not be bands but the hemispheres of daylight and darkness which the earth wears like crowns day and night. The crowns of unmixed fire on this account are narrower than those of night because in the daytime we cannot see as far above us as we can at night. In the midst of these crowns is the demon goddess, and since they alternate in time and are not worn simultaneously, we must look for the place in their midst not at some putative center in space but at the moments and places of their alternations. These crowns meet at dawn in the east, in the west at dusk, and the demon goddess between them is there-fore likely to be an evening and morning star, Mercury or Venus. The crowns which are mixed of night and fire as Aetius tells us are the times of halflight at dawn and dusk and again in the very midst of these is this star. The portion of flame which Night describes as sent between the crowns is the red sky of sunset and dawning. In the midst of the crowns the demon goddess governs all things and she governs this portion of flame. Let us suppose then that the gates of the opening lines are the gates of dawn and dusk and that Justice is the demon goddess who is Venus.

The goddess in the palace is Night, and Justice is the Evening and Morning Star. Then the rider in the chariot is not only the poet but also, in some way, the sun. The chariot must be the chariot of the sun. Then the chariot ride to the palace would be a symbol of the meeting of Day with Night in the palace of Night. This possibility does not square even with the Orphic tradition that Night and Apollo shared an oracle, since this does not require their being together in the one place at the one time. Hesiod in his *Theogony* says that Night and Day are never together in the palace of Night but the one always goes out as the other comes in.[2] This makes sense,

1. H. Diels and W. Kranz, *Die Fragmente der Vorsokratiker* (Dublin/Zürick: Weidmann, 1972) Aetius, ii, 7, 7.224.

2. Hesiod, *Theogony*, 135.

but it does not provide a reason for calling the palace the palace of Night. It seems to me inescapable that Parmenides describes in his opening lines a meeting of the sun with Night in the palace of Night, wherever that may have been. The chariot is indeed a symbol of the solar vehicle: its axle burns in the boxes; it is escorted by the daughters of the sun; it is pulled by immortal horses; it travels across every city on the road of a demon goddess; its rider is addressed by the goddess in her palace as young man, a conventional way of addressing a god. The rider crosses the stone threshold of the gates of the paths of Night and Day. Since the rider is not Night he is more than likely Day. These gates are ethereal, high as heaven. The chariot traverses vast spaces at the greatest speed. There is a feeling stronger than exhilaration, of exaltation, of being lifted high above the world and of being directly controlled by divine powers. Even the piping shriek of the axle in the boxes recalls the Pythagorean music of the spheres.

The rider in the sun chariot reaches and enters the palace of Night. He enters in the west at dusk just as we go into the darkness then. This is an image drawn from experience, not from an ignorant belief that the sun disappears at nightfall altogether. The image describes the apparent movements of the sun according to an animal rather than abstract astronomy. In the same way the sun is to be imagined as leaving the palace of Night in the east at daybreak when he comes into the light as we do. We are told that the maidens who escort the chariot to the palace of Night have first left that palace for the light. At this time too we should imagine them to have been escorting the chariot. These maidens the poet calls the daughters of the sun and they escort the sun from east to west and into the palace of Night. They are the stars of whichever zodiacal constellation the sun happens to be in. The accompaniment of the sun by these stars is the first datum of astronomy; the sun's regression round them is the annual cycle. There is, perhaps, another image of the stars in the detail of the palace doors, the two posts of which are studded with bolts and nails. By other poets the stars are called crystal nails. Here they adorn the bronze doorposts of the palace of Night.

Justice is the Evening and Morning Star. She is where the sun goes down. The sun leaves the palace of Night at dawn in the east and then Justice needs her other key to let him out. This is consistent with a maxim of Heraclitus, Parmenides' contemporary, which describes the work of Justice:

> The sun will not overstep his measures. Otherwise the Furies, ministers of Justice, will find him out.[1]

A Justice as grim as Parmenides' lady of punishments, and who also plays a part in the movements of the sun. The measures of the sun are its limits in relation to night, so Justice compels the sun to go down according to Heraclitus. She is assisted by the Furies who were born from the blood shed at the castration of the sky god Uranus, when Aphrodite was formed from his bloody member. She does the same work according to Heraclitus and Parmenides but there is a marked difference between their accounts. For Heraclitus Justice is a power of restraint, curbing the sun's tendency to excess. For Parmenides she is, with Themis, the driving force of the sun's journey. Her punishment, we feel, would be to keep him outside the palace of Night and not a means of forcing him in. Nonetheless we are bound to consider here the possible identity of Heraclitus' Furies with Parmenides' Sun Maidens.

The demon goddess between the crowns of night and fire rules also over hateful birth and sexual intercourse, sending female to mate with male and male to female (XII). This is confirmation that the demon goddess Justice is a form of Aphrodite. The lines which describe her place between the crowns redescribe the features of the chariot ride; so do the lines following them about love and birth. Now Aphrodite with the keys controls events from the mound of Venus above the female gates, the palace of Night is the female body and the chariot the male who enters her in the act of love. We are not yet done with the passing of all things through all. Night's description of how the demon goddess sends female to male and male to female redescribes the interactions of maidens and rider as the exciting reciprocities of sex. We see why the horses are straining at the car and why the maidens push back their veils from their faces with their hands. The chariot and horses which were the sun a

---

1. *Ancilla to the Presocratic Philosophers*, op. cit., Heraclitus, xcɪv.

moment ago are now the rampant phallus. Justice is cruel indeed who may punish the rider with exclusion from the palace, but the serenade of the maidens urges her to relent. The doors open, the chasm appears and on down the road to where the goddess within takes the rider's right hand in her hand. The rest of the poem is a giving of wisdom from the womb, the underground cave of Pythia or Sibyl, in the place and at the moment of union between male and female. The closest word to Delphi in Greek is *delphys*, womb. On this account of the opening lines the being which Night describes is the newly conceived, like the mass of a well rounded sphere stretching equally in all directions from the center (VIII).

There are symbols of birth too in the chariot ride. The maidens who leave the palace of Night are the stars of the zodiacal ascendant, a variable to be determined in the casting of a nativity. Their pushing of the veils from their faces was erotic a moment ago, but from this point of view represents the first opening of a new born baby's eyes. Empedocles tells how the creature in the womb does not behold the swift limbs of the sun nor shaggy earth nor sea, but is a rounded sphere enjoying a circular solitude.[1] To be born out of this condition is hateful indeed (XII). Here are the gates of the paths of Night and Day through which we come into the light of day from the darkness of the womb. Justice herself pushes out the bar from these gates, a bar which is rounded at the top like a date or acorn, but here like the head of a child being pushed out first as the child is born. How far the other mechanical details of these lines were suggestive to the poet of love and birth it is hard to know. Keys, axle, doorposts, sockets and boxes may all represent these same realities of flesh and blood.

Love, we read, was the very first of all the gods to be devised (XIII). This tells us what we already know, that the poem itself begins with an account of love. Who devised love we are told by Plutarch who quotes this fragment. Love is Eros and he is devised by his mother Aphrodite.[2] Love is devised, a work of cunning like the curiously embroidered belt which Aphrodite lends to Hera in

---

1. *Ancilla to the Presocratic Philosophers*, op. cit., Empedocles, frag. XXVII.
2. Plutarch, *Moralia*, v, 'Dialogue of Love', 756E.

the Iliad.[1] With this belt Hera hopes to seduce her husband Zeus. In its intricate stitching, Aphrodite tells Hera, is all power over love and desire. Parmenides' first lines are a device of just this kind, wrought to the last detail, its power learnt from a following through of each of the twining threads, as chariot ride or as sunset or as the act of love. Sappho calls Aphrodite a weaver of tricks.[2] There are also the lines about the mingling of the wandering limbs when mind has come to people (XVI), and the lines about how man and woman mix the seeds of love (XVIII).

* * *

The name Venus was given by the ancient Romans both to the brightest of the visible stars after the sun and moon, and to their goddess of love. The character of this goddess they mostly borrowed from the Greeks who called her Aphrodite and the planet the star of Aphrodite. Why did the ancients name that planet and that goddess by the same name? It is well known that they gave the names of their gods and goddesses to the stars, and why they did so in general is a question in itself. But here I am asking why the name of this goddess for this star. It is true also that they venerated physical love to a degree which later generations have not and certainly the planet Venus is remarkable for its size and brilliance. The grandeur of the star reflects the centrality of the cult. But their reason for so naming the star is more complex than this.

It is not the brilliance alone but the times and places of its appearing which explain the planet's name. It appears, when it appears, at dawn in the east or at dusk in the west. We put this now by saying that Venus is a planet which closely circles the sun and always appears close to the sun to us who are less close. In this connection some ancient and modern names for the planet are significant. There are two lines of Homer describing the flashing point of Achilles' spear:

---

1. *Iliad,* xiv, 220.
2. *The Pageant of Greece,* ed. R. W. Livingstone (Oxford: Oxford University Press, 1935), Sappho, I, 2.

like a star among the stars at the milking time of night, Hesperus, the
most beautiful star which stands in the sky.[1]

The name Hesperus is related to the name for evening in Greek,
in Latin it is Vesper and in English we call Venus the Evening Star.
Likewise we call it the Morning Star, in Greek *Phosphoros*, in Latin
Lucifer, the bringer of light. These names indicate the character the
ancients discerned in this planet and our own names show we are
with them in this.

There is a question, however, whether or not the ancients identified the Morning and Evening Stars as one star from the beginning,
that is, from the time of Homer, and again whether they had always
called this star by the name of the goddess of love. Writers in
Roman times attributed to both Pythagoras and Parmenides the
discovery that the Evening and Morning Stars are the same star[2]
and both were known to have lived much later than Homer. It is not
until later still that we find the first instance of the star's actually
being called the star of Aphrodite.[3] Nonetheless the doctrine which
I expound here is, I believe, from earliest times. The energy and
imagination which first established Aphrodite as a goddess must
have been incomparably wiser and more complete than any which
shared in it after. It is hard to understand how anyone could have
failed to recognize one of the brightest lights in the sky wherever
and whenever it appeared. It is impossible to understand how
Homer who instituted the worship of Aphrodite could have known
her less well than those who succeeded him in a worship already
established. Progress in understanding may occur with theories but
not with visions of divinity, with scripture or with rites. They are
whole from the beginning.

Let us return to the question. Why is that star called Venus, goddess of love? Because Venus is the Evening and Morning Star and
presides over the movements of day and night at dawn and dusk
just as Venus, the goddess of love, presides over the movements of
male and female in the acts of love and birth. But is it really the

---

1. *Iliad*, XXII, 317–318.
2. Diogenes Laertius, op. cit., IX, 23.
3. *Plato, Collected Dialogues*, op. cit., 'Epinomis', 987B.

same power which presides over such different operations? That it is the same power, that day and night and male and female are similar pairs, is the first principle of this doctrine. The realization of this formed the character of Aphrodite and the scale of the realization made them call her a goddess. First of all she is that one in whom the junctions of day and night on the one hand and the junctions of male and female on the other are just two aspects of the same one thing, herself. The discovery of a single cause underlying two manifestations as enormous and seemingly diverse as the heavens and sexuality, this discovery transformed itself into the worship of the person of a goddess in whom as at a point the heavenly and the animal intersected. As we shall see, this discovery lies very close to much we say and feel now.

We have answered the question and explained the meanings in the word Venus. Many people know Venus to be the name at once of a planet and of the Roman goddess of love so we have made explicit a nexus widely felt. The word Venus is an ancient treasure which has suffered many vicissitudes but has at last come down to us to remind us of a great truth: the analogy between day and night and male and female. The power of the analogy is, we may say, absolute. Everything we find in the sky related to the planet Venus we find in our own bodies related to love. The proper marrying of the astral and sexual realms will in what follows sometimes seem absurdly literal. But it is unreasonable to expect that all the ways of ancient thought will immediately appeal in this age. We have changed our minds since then.

How did those Greeks align day and night with male and female? By discovering correspondences between the lights in the sky and the reproductive organs of bodies. They identified day with male and night with female, reasoning that the female reproductive organs are essentially dark, hidden, while the male organs are exposed. They thought of night as a dark receptacle bounded by the fixed stars, as the kindly womb of night out of which the sun is born every day and to which it returns every night. The sun, the single star of day, they identified with the phallus and especially with its head which in color and shape most resembles the sun. Night is a receptacle, day a point which moves through the space contained by night. We may think of night in the figure of a circumference from which the radii move inwards toward the center and of the sun as a center from which the rays move outwards toward the circumference. In this way day and night are a pair like male and female, alternating pulses essential to each other, the one the collapsing of light into a single center, the other the dispersal of light to the outermost limits of the sky.

The sun is the phallus of the sky. There is an old story told by Hesiod, one of the earliest Greek poets, in his *Birth of the Gods*[1]: how once upon a time the earth goddess Gaea groaned under the oppression of her lord, the sky god Uranus. For he constantly impregnated her, yet would not allow her to release the children so conceived into the upper world. But she freed one child, a son called Cronus, to whom she gave a saw toothed stone and told him to castrate his father. And he did and from the drops of blood so shed upon the earth the Furies sprang who avenge crimes against fathers. But the genitals of Uranus fell into the sea and there after some time in the foam around them they were transformed into the body of Aphrodite who rising fully formed came to the island of Cyprus. This story of the goddess' birth gave her several names: Urania, Aphrodite of the sky; Cypris, from the island she reached; and the name Aphrodite itself which Hesiod takes to mean 'of the foam'.

This is one of the most delightful of ancient stories. Its wicked savagery masks a most pleasing spectacle. For the phallus of the sky god is the sun and the saw toothed stone which Cronus takes to it is the jagged ridge of the western horizon, and the very moment of excision is that lovely moment as the sun sinks slowly out of sight beneath the world's edge. Then the western sky is reddened by the bloody member till in the midst of it appears the Evening Star. It is a very striking story to describe a simple thing. But how better describe the conditions under which this star appears, if the purpose is to impress upon the mind the analogy between our sexual organs and the lights and colors of the sky?

Sun is the masculine phallus, night the feminine womb. There are gates through which the sun passes. Through those in the east

---

1. *Hesiod, The Homeric Hymns and Homerica,* op. cit., 'Theogony', 154–206.

the sun emerges as from the womb of night. Through those in the west he reenters that womb. As the ancients identified male with day and female with night, they identified dawn with birth and dusk with love. For they thought that just as the phallus penetrated head first into the feminine darkness in the act of love, so at birth the head of the child first emerged from that darkness, moving in the opposite direction. This depends upon a further identification between the head of the phallus and the head of the body. The two heads are similarly shaped and both are at the top when erect. Both too are the most sensitive parts of their wholes. A most sensitive organ in the head is the eye and the eye corresponds to the sun which is the eye of the day. But the analogies between dusk and love and dawn and birth are more complex and imaginative than those between day and male and night and female. For the sun in a day passes through two sets of gates which are opposite each other and widely apart, and it passes through them going in the same direction. But the gates of love and birth are the one set of gates, the one part of the female body, and love and birth are opposites by virtue of their opposite motions in the one place.

Or are we to say that here also there are two sets of gates, the inner and outer, or greater and lesser lips? Or are there other gates still? For the analogy between the two realms to be absolute, everything we find in the sky related to the planet Venus we must find in our bodies related to love. Certainly the goddess herself is in her rightful place on all four occasions. For as star she is seen to stand above the sun after he has sunk beneath the western horizon, as she stands above him before he rises in the east. Likewise she presides over the union of male and female on the mound which bears her name, the mound of Venus above the female gates, and she stands in the same place relative to the act of birth.

The moon too has a part in the dance. Though the moon alone can sometimes be seen in daytime with the sun, its light is much brighter at night. The animal analogue to the moon is therefore to be found in the female reproductive system. The moon waxes and wanes and the full cycle of its growth and decay takes a lunar month. This is the period of the human female's menstrual cycle as it turns from infertility to fertility and back again. Because the moon is larger than the other stars in the night sky, and because its light is borrowed from the sun, the ancients thought the moon to be the sun of night. Its material analogue in the female reproductive system is the rudimentary female phallus which is called the clitoris. This sensitive part of the female anatomy corresponds to the phallus of the male as the moon does to sun.

There is an old story how once Zeus, king of the gods, begat a child by Alcmene, a mortal woman and wife of Amphitryon.[1] The child was Heracles. Father Zeus wished him to be strong beyond mortals and so placed him unannounced upon the bosom of his own wife, the jealous Hera. The child gave suck and Hera's milk came out so vigorously that some of it flew up and made the stars. Tintoretto made a painting of this moment. The stars are the divine milk of the archetypal nursing mother, cow-eyed Hera. The idea has come down to us in our description of some stars as the Milky Way. Our word galaxy is a word derived from the Greek word for milk. The stars correspond to the female breasts as the sun to the phallus, night to the womb, and moon to female period and clitoris.

Day and night, male and female are alike in their pairings more intimately yet, as the cult of the goddess reveals. For the flower sacred to Aphrodite is the red rose and the color rose red. In our time we talk of red light districts and pornographers discuss exposure of the pink. Red are our lips and nipples and our most private parts, reddest when closest to the consummation of love. Then they distend and bloom with the redness of blood. This is the blood shed by Cronus when he castrates his father Uranus and the phallus falls into the sea. The phallus is the sun which froths and colors the white cloud and the star of the goddess appear standing over its redness.

There is on this theme a poem by the Lesbian poetess Sappho in which she addresses a companion concerning a mutual friend of theirs who has married and gone to live in the Lydian city of Sardis:

> From Sardis her thoughts often turn hither. When we lived together, she held you ever as a goddess and loved your singing above all. Now she shines among the women of Lydia, as the rosy fingered moon, when the sun sets, shines brighter than all the stars. The light falls on

1. Diodorus Siculus, op. cit., IV, 9.

the salt sea and on the fields deep in flowers. The dew descends in beauty, and the roses are in bloom and the clovers and flowering grasses. Arignota wanders up and down and thinks of gentle Atthis, and her thoughts are heavy with longing and her heart with distress. She calls aloud for us to come to her, but Night with the thousand eyes does not carry the words across the sea, and we cannot hear.[1]

Even without what I have said, it is hard to read this and not feel that it describes in part the growing heat of feminine sexuality. The roses bloom on in the moonlight, the dew falls on the clovers and flowering grasses. This dew is the last of the analogies between day and night and male and female with which we are concerned here. On the night of the last full moon of the year, at midsummer, Athenian maidens ritually gathered dew in the gardens of Aphrodite Urania.[2]

---

1. *The Pageant of Greece,* op. cit., Sappho, xcvi.
2. Harrison, *Themis,* op. cit., p173.

3

NIGHT

PARMENIDES' POEM begins with an introduction which describes in the first person how the poet journeys by chariot to the house of a goddess, how the goddess receives him in her house and tells him the rest of the poem. This story is told with details which excite many interpretations. So far we have considered it under the aspects of political vindication, oracular consultation, dusk and dawn, love and birth. But whatever else the story may describe, everyone would agree that it describes this at least, the course of a philosophical experience in terms of the physical experience of a chariot ride. Parmenides did not write very much. The little we have is a large part of it and of this little the opening lines are themselves a large part. So we cannot readily suppose that Parmenides denied the reality of such things as philosophical revelations or the physical experience of a chariot ride. In any case someone who can describe a chariot ride as vividly as Parmenides does is not likely to have undervalued his everyday experiences. But in fact many of Parmenides' interpreters, from early times and recently, do suppose that he denies in his poem the reality of the physical world. Parmenides' name is now generally associated with the theory that the physical world has no reality.[1] He is regarded as the originator of this theory. But we may not readily suppose that Parmenides denied the reality of the physical world, for this is what his opening lines describe.

The reason scholars have for supposing that Parmenides denied the reality of the physical world is his account of being, the central passage of his poem. It is true that this being is on first acquaintance

---

1. Plutarch, *Moralia,* 'Against Colotes', 1114D.

peculiarly cold and inhuman. Immobile, perfect, eternal, unitary, integral, and homogeneous, its identity seems to consist largely in its dissimilarity from anything we might recognize from the physical world, except perhaps space. To this being, Night actually says, mortals ascribe all names in the belief that they are true, coming into being and destruction, being and not being, changing of place and color (VIII). But her implication is that all these names except being are wrongly ascribed to being, which she has just proved to be quite otherwise. And so scholars deduce that the lively chariot ride can have nothing to do with the being which Parmenides describes after it, even though the chariot ride is his means of reaching the understanding of that being. To save Parmenides from one contradiction they involve him in another. It is remarkable how being is cold and abstract while the chariot ride is passionate and physical. It is quite possible that Parmenides does begin his poem misleadingly and did not seriously intend as much of life as the opening lines convey. But we may adopt this interpretation only after considering and rejecting the alternative, that Parmenides thought he could have both worlds, being and the physical world. And this, it seems to me, is what Parmenides and the goddess Night are arguing, whether we judge the argument successful or not. For Night tells the rider that he must discover all things, both the unshaken heart of well rounded truth and also the opinions of mortals. And then she tells him how these mortal opinions, the things that seem, yet had to be, all of them passing through everything, Night makes a distinction here between being and seeming and then instantly confounds the distinction by claiming that the things which seem assuredly are. She is far from believing that the things which seem have no reality, but she does claim, I think, that the things which seem only have reality insofar as they all pass through everything. For then, of course, the things which seem are homogeneous as being is.

Night tells the rider he is to learn of both being and seeming. She relates seeming to being at the beginning of her argument and affirms its reality. The things that seem are in general our human experiences as creatures living the world, but more particularly they are the experiences we have as we read the opening lines of the poem. Their all passing through everything is what the structure of the opening lines is intended to convey. All the people, animals and objects in the opening lines are multivalent, merging into each other in their indefiniteness, brought all together at the palace of Night, turning as we read into the many different things each symbolises. This is the first point the poet has to make about reality. Though we think of each physical object, event and experience as distinct from all others, it is not so. This is what the rider and the reader have to learn from the chariot ride. The passing of all the things which seem through each other and everything else, the homogeneity of seeming, is realized implicitly in these lines. Parmenides finds a way to express this interpenetration of things by a special artefact of language, a difficult task since language is what makes the distinctions he has to confound. In this way the opening lines introduce the account of being by illustrating the nature of the thought or thinking which the account of being describes. For being is thinking, Night says (III). The story of the chariot ride is not inconsistent with the account of being, it is our best means of understanding what being is. Being is the thinking which the opening lines enable us to perceive clearly in its inward unity and homogeneity. From this point of view the account of being, the central and most famous portion of Parmenides' poem, is the organizing principle of the opening lines. The goddess always returns to her starting point.

Parmenides' poem opens with a mad onward rush of passion by which he is carried along out of control to his passion's fullest satisfaction. In a single moment he becomes aware of a range of different experiences and memories all at once. His mind begins to operate consciously at many different levels at the same time. This simultaneous multiplicity of experiences is communicated to us who read the introduction by the number of different meanings suggested by almost every word the poet uses. It is this mental rush which the din and speed of the chariot ride best represent. The poet's mental hyperactivity occasions a corresponding withdrawal from attention to the senses, a withdrawal so intense that it is almost sleep.[1] This is signified by the rider's entry into the palace of

1. Diels and Kranz, op. cit., Simplicius, *Physics*, 39, p. 22.

Night in which the goddess dwells. In this state of withdrawal the poet is aware of the operations of his own thinking only, an awareness as broad as his withdrawal from attention to the outer world is complete. He understands how these many different experiences which now crowd upon him comprise his thinking, and how each experience which he conceives is shot through with every other. He understands that his thinking is his being and that each of its elements is necessary to the whole and inseparable from it. He understands that everything he had previously considered apart from himself is in fact present to him so far as he now conceives it. The goddess bids him behold how things which are absent are firmly present to his mind nonetheless (IV). He understands that his past and future exist only insofar as he now conceives of them in the present, that his being or thinking neither was nor will be but is now all together, one and continuous (VIII). And he understands by the peculiar logic of being that his being has no beginning nor end, is unchanging and immobile. In short, he experiences the supreme moment of self consciousness, the clearest and most direct of all experiences, and he sees how his own present nature satisfies his every desire for immortality, perfection and unity. He becomes aware of what he is always and thinks his new found awareness divine.

Sleep is paradise, every night reminds us of timeless peace. Sleep also is now and forever, there is nowhere to go, we are here. Eternity is as close to us as when we were last asleep, when the great world was no threat to our minute existences but was itself held safely and whole within the heart of each of us. Here is the all inclusive sphere of being, and the unshaken heart of well rounded truth. Parmenides did not differ from us constitutionally, the difference lies in his knowledge of the self we share with him. He knew how the energies of things pulse through a larger circuit than we know awake and alive, through darkness, sleep and death. Like Orpheus,[1] Odysseus[2] and Aeneas[3] he passed while yet himself within the halls of sleep

1. Diodorus Siculus, op. cit., IV, 25.
2. *Odyssey,* XI.
3. *Aeneid,* VI.

and death and found out the other side. This is the work of initiation and frees the initiate from the fear of death. But the chariot
ride reveals our sleeping selves and the nature of waking consciousness too. The presence of all experiences which is characteristic of
the sleeping phase is also the condition of the mind awake. The
rider is made to see that any experience or seeming whatsoever is
what it is to its subject by virtue of its relations to all his other experiences. Each experience anyone has is chosen, censored, valued and
shaded in accordance with its place in the scheme of all their experiences. The entire scheme is present, but hidden, at every moment of
our waking life as the stars are still above us in daytime though
obscured by the sun. This is the self Parmenides comes to know,
which he calls being and to which he attributes the perfections of
divinity. It is as though the sun were to shine in mid heaven at midnight and reveal to his eyes the innermost secrets of the stars. Or the
stars shone out brightly in daytime as they escorted the sun on his
journey.

What makes it happen, this mingling of the day and night of the
mind, how is the waking self extruded into the darkness of sleep?
What enables the rider to remain aware as himself after he has
passed the gates of sleep which normally exclude the waking consciousness? We are told only that he is brought to the palace of
Night by immortal horses which know their own way and that he is
admitted to the palace by Justice who is persuaded by the daughters
of the sun. The first line of the poem tells how these horses take the
rider as far as the heart could desire. His journey is to the goal of
life, the ultimate object of desire. The discovery of the greater self is
the only possible end to the soul's search since any other would be
insecurely attained in comparison. But being is irremovable from
us, the unshaken and inviolate self of the soul, its inseparable and
hidden joy. All times and places, experienced and imagined, are
comprised in the sphere of being. The sphere itself is neither in time
nor space but prior to both. No state nor condition of anything
whatsoever is any more part of the sphere of being than its opposite
or any other state or condition. Being itself is beyond all opposition
and beyond process and change. This place beyond place and time
beyond time is imagined in the figure of a sphere which comprises

all places and times but is itself at none of them. Night also calls it the unshaken heart of well rounded truth, the center of a sphere equidistant from all points on its surface, yet including them all within itself like a seed. Nothing is ever forgotten. We carry the world in memory at all times and suddenly all of it can be opened to us at once as when the whole of life flashes before the eyes of someone nearly drowned. To see for a moment what one is always. The Greek word *aletheia*, which is translated as truth in English, literally means unforgetfulness, and for mortals the only way to that is through some such recollection as Parmenides describes.

Truth is the recovery of being, the revelation of what has always been there, not the finding of something previously unknown. Being is prior to any experience because it is a precondition of any experience. It is older than seeming as Night is older than Day according to Hesiod.[1] The stars are always above us, but the sun is not. Both sexes are carried first in the womb. Likewise the sleeping self conditions waking life. The things that seem assuredly are, so far as they are known in being. They merely seem so far as they are not so known. Each thing that seems is what it seems by being discriminated from everything else, but in being there are no discriminations. It is hard to see how the rider would be capable, in the state he describes, of having any experience at all. Any experience he had would be submerged, as it were, by its component associations and by their associations, all realized as clearly in the mind as the immediate experience. How would a person live to whom the present he experienced was no closer than the past he remembered and the future he anticipated? In supposing all his experiences equidistant from his true center, and standing there, he would be unshakable and inviolate, in an equilibrium beyond which there could be nothing to disturb it. By a long process of reflection on his experiences he would catch up with them and perceive their influence on his daily understanding. He would know the world on the instant in terms of them all. He would know the mind which knew as well as what it knew. The Pythagoreans made a practice of recalling the

---

1. *Hesiod, The Homeric Hymns and Homerica,* op. cit., 'Theogony', 124.

events of the day just past before rising in the morning.[1] This may have been one of the lessons Ameinias taught Parmenides to prepare him for the lesson of Night.

The psychological interpretation of Parmenides' opening lines follows upon the astronomical and biological. The psychological interpretation has to be given after other interpretations because it describes the combined effect of several interpretations working upon the mind at once. It is an interpretation of the second order, the other interpretations are of the first. The procedure of the astronomical and biological interpretations is to make the chariot ride stand for other events of a physical nature, and we note the physical similarities between the chariot ride and these other events. But the psychological interpretation has the chariot ride stand for a mental phenomenon. Mental phenomena cannot be pointed out as readily as physical ones, especially not such a mental phenomenon as I take Parmenides to mean. So the opening lines enact this phenomenon in the way they are written and we are enabled to see this by what else we are told of the chariot ride in the last fragments of the poem. Poems which describe their own effect upon the reader are not uncommon. This poem does not give us a metaphor to stand for things, events or even mental phenomena, but a metaphor to stand for the processes of metaphor. The single metaphor of the chariot ride first stands for various events and then for the variety of them.

But though the opening and concluding fragments of the poem illuminate the being which is the subject of its central fragments, they do not exhaust it. Being is manifested in the astronomical, biological and psychological realms but is itself beyond them all. The order or pattern found in the three realms is the same because being is prior to all three and informs all three. If however we attempt to conceive of being as beyond the world, it must be by a process of elimination, till being is neither the sphere of the stars nor the womb nor even the heart's thinking, but something else altogether with its own laws and character. It is this which determines but is

---

1. *The Pythagorean Sourcebook and Library,* op. cit., Iamblichus, 'Pythagorean Life', 165.

not determined by its manifestations. The world is incomprehensible except so far as we comprehend it in terms of being, but being cannot be known in its own place from our understanding of the world be-cause it is beyond the world.

This makes the being of Parmenides sound very hard to understand, and it is if we think of understanding as something to be demonstrated to other people or to ourselves. But at another level supermundane being is not hard to understand at all. It is like a magnet which draws the mind to itself as the chariot is drawn to the palace. So it is that readers of Parmenides often feel they do understand something of the central fragments whether or not they think they understand the first and last fragments. The account of being touches us because it is self-contained and self-determining. It is learnt from a contemplation of being, not by inference from its manifestations in the world. Being must first be seen in its own place before it can be recognized in phenomena. Nor is being to be understood by a process of deduction. In Parmenides' system there is no principle prior to being from which being is deduced. Nor does the goddess derive the attributes of being by a conceptual analysis. The hearing of her discourse is a more tremendous experience for the rider than this. Night's account of being is a demonstration, since in the supermundane realm to see that a thing is so is the same as to see that it must be so. For there is nothing else of it to see. The dialectic of Night is a way of transmitting the inward certainty of a revelation, not of proving something true to one who does not know it.

* * *

We have seen what is meant by calling the goddess of love and the planet Venus by the same name. But in fact the analogy which is represented by Venus and Aphrodite is more complex still, since it concerns not two distinct dimensions but a third also, as distinct from each of these two as they are from each other. That Aphrodite is a goddess presiding over three realms and not two is what is meant by the attendance upon her of three Graces.[1] These ladies are

1. *Hesiod, The Homeric Hymns and Homerica,* op. cit., 'Theogony', 509.

called Aglaea, Thalia and Euphrosyne and they attend to the beauty of their mistress. Now Aglaea means Shining One, She who shines, and signifies that aspect of Aphrodite which has to do with the celestial luminaries. And Thalia means Flourishing One, She who blooms, and signifies that aspect of Aphrodite which has to do with generation and birth. With these twin aspects of the goddess we have already dealt. But the third Grace is called Euphrosyne and her name means Well-Mindedness or Joy. Her presence in this trinity signifies the power of Aphrodite over the operations of mind also.

The notion that Aphrodite plays a part in the operations of mind is less familiar to us than that she plays a part among the stars and in the act of love. But if we are to follow the ancient theology, the gods and goddesses are from the very first to be considered as intellectual entities. Aphrodite's part in the operations of mind is not only unexceptional but necessary since the operations of mind are precisely what intellect looks to. Since Aphrodite is the name of an important goddess, she must also be a power of the soul. And in the same way. As planet, Venus presides over the most momentous events in the heavens, the alternations from day to night and from night to day. She must therefore bear an equal part in the events of the soul. And this part must correspond as fully to the part she plays in the acts of love and birth. In this way we shall establish the trinity of the Graces.

Day is to being awake as night is to being asleep; and as dawn is to waking up so is dusk to going to sleep. Aphrodite or Justice is the power who presides over all these, since in her, by virtue of this pattern, the astronomical and psychological realms are at one. I call dusk and dawn the most momentous events in the heavens, and these two with a day and a night constitute a primary cycle of the heavens. Likewise I call waking and being awake, going to sleep and being asleep a primary cycle of the soul. As for the analogy between the two realms at this point. I do not know what to think of it. Why are we awake during the day and asleep at night? Have we adapted to the cycles of heaven or should we posit the primordial establishment of a harmony between the astral and the psychic? And what is the force of the analogy? It is too obvious to think about and impossible anyway because nobody knows what sleep is. And this is true,

we have no reliable data concerning one of our terms and so we can
no longer use the method of pointing out one thing and then
another to make the connections. Instead we must use the method
of hypothesis and by assuming the connections are there, see what
we can make of the invisible psychic events by extrapolation from
the worlds of star and sex. For we must remember always to look for
the links to the third realm as well.

Being awake is like daytime, and not just contemporaneous with
it, only if we suppose an analogy between the sun and waking consciousness. This is not hard to make out. The sun is lofty, solitary
and brilliant. It goes on high above the world which it reveals to our
sight, by the one act revealing its subjects and establishing dominion
over them. The stars, though higher, do not do this. The sun puts
the stars to rout with its obscuring brilliance and stands jealous and
solitary in the sky, allowing none other to enter its presence except
the wan moon. Though the world itself shines in the sun's radiance
it does not shine as the sun shines. All this we may find in the waking consciousness which is characteristically the state in which we
use our senses. The organs of our senses are points at the centers of
fields of awareness which constitute normal waking consciousness.
When we are awake, each thinks of himself as one thing and of
everything else as being somehow outside himself, objects to subject, known to knower. How far this otherness is felt is another matter, but the distinction between ourselves and others is the basis of
our individual animal self awareness. This corresponds to the difference between sun and world as things visible. The seat of most
senses is the head. As the sun is the eye of day, so the whole head is
the sun of waking consciousness, the radiating center. Every morning we raise our heads from the pillow and rise as the sun rises above
the horizon. We imitate the sun's unsupported elevation by this rais-
ing of heads on the props of bodies. Since, when we stand straight,
we look out, not down, we create for ourselves that separated
dominion over the world which is char-acteristic of the sun. And
our arising in this way in the morning repeats not only the rising of
the sun but also the rising of the phallus, its standing out from the
body. A human is a sun on a stick.

But we are also sleepers when like the sun we lower our heads once more to the pillow and go into the darkness. This transformation is more than just the vanishing of the visible world of objects, it is a turning of the very greatest wheel in the machinery of the mind. At this point we must recall that we are in fact presently awake, I writing and you reading, and that neither of us is aware now of doing much of either when we are asleep. The problem which any account of sleep must overcome is insurmountable: there is no other way of explaining the phase of sleep except in terms of waking experience to us who are awake. This problem is like the mutual incomprehensibility of the sexes. We must work with the two analogies of night sky and female body. I suggested before that night is a circumference of stars from which the radii move inward toward the center, and the sun is a center from which the rays move outward toward the circumference. This figure enabled an analogy between male and female and day and night, and it is also a figure of being awake and being asleep. The radiating center is the waking consciousness which conceives of the world as outside and at a distance from itself, and the concentration of the circumference is a figure of how in sleep the world is within. The inwardness of sleep is suggested also by the completeness of the astral cover. Unlike the day there are stars everywhere in the night sky and the world is held within them all as the soul is sealed in sleep and as the child is carried in its mother's womb.

Surrounding stars against centered sun. The sun is what it is by virtue of its distinctness from the world which it illumines, and in this way corresponds to the waking self. The sun is always at a different point at each moment. This is true of any one star also but it is not true of all the stars together. There are so many of them everywhere in the night sky that it seems always and equally full of stars. There is a constancy in the night sky but the sky of day is continually being changed by the motion of the sun. This constancy of night represents the timelessness of sleep as the movement of the sun represents the spatio-temporal continuity of waking consciousness. In sleep there is no other and therefore no one; when the world has vanished into darkness and there are only the stars, there is neither up nor down; among the stars there is no then nor now. The conditions of time and space are removed from our consciousness; we ourselves as selves are removed from consciousness; consciousness itself disappears with the collapse of the division between subject and object, and we enter into our other life.

There is a phase intermediate between the sleep which is represented by the fixed stars and the waking consciousness of the sun. I mean, of course, that phase of the mind represented by the moon. The moon is the sun of night and banishes in its fullness the stars nearest it. At the same time it lights the world of objects on earth. But its light is dim, the objects scarcely seen and all the other stars shine out around it. These are the properties of the dreaming self, the pale center of the sleeping life. The dreaming self perceives its phantoms somewhat as the waking self perceives external objects and we recount our dreams as though they were waking experiences. We are their subjects but awake we know that dreams occur within us, as the moon is of the night. Sometimes we hover between the two, dreaming a dream and at the same time knowing that it is a dream, and sometimes we feel a terrible impotence when our dream self moves slowly on to a foreseen destruction, as though, not under our control, it could be fully if we knew but how. Like the moon the dreaming self is dim where the sun of consciousness is brilliant. Here the center is too strong to permit consideration of itself. When awake we cannot quite imagine that the waking world is within us as we can, dreaming, of our dream. Somewhere between dreaming sleep and waking are the gates of horn and ivory through which true and false dreams come into the world, according to the Roman poet Virgil.[1] Over these gates too we may suppose Venus to preside. And it is well known that madness and delusions of the self have been believed to wax and wane in periodic cycles of the lunar month.

The moon corresponds to the reflection of the waking consciousness which manifests itself in sleep, the dream self. Likewise moon and dream self correspond to the rudimentary phallus of the female body, the clitoris, and to the female period. The world which the moon reveals by reflecting the light of the sun is the same world as the world of daylight. Similarly the world experienced by the dream self is the world of waking consciousness so far as dreams are

1. *Aeneid,* VI, 892.

memories of waking life. But the moon differs from the sun in not obscuring the stars completely and we may take it that whatever in dreams is not memory is to be attributed to whatever the stars symbolize in the operations of the mind. The stars are remarkable in the way they appear at dusk and disappear at dawn, because their appearance and disappearance correspond exactly to the disappearance and appearance of the world of daytime objects. This is, I think, more obvious at dawn when the successive dwindling and disappearance first of the smaller and then of the larger stars parallels the slow solidification of the shadowy and finally fully colored world we know by day. This phenomenon points to an equivalence between the things of this world and these celestial bodies. The stars are the things of this world laid up in heaven, not merely the signs of the seasons but the very things of this world in a finer form. The names given to constellations indicate a widespread tendency to this reasonable belief for they are the names of things in this world and not just names for the constellations alone.

The fixed stars are self luminous. Unlike the things of this world they do not require the external agency of the sun in order to appear. Indeed their appearance requires its withdrawal. This self luminosity is part of the reason why the ancients thought them to be divine and it gives a clue to their counterparts in the phase of sleep. For we must suppose that there are in sleep also entities which are seen, as it were, by their own light, and which correspond to the objects of our waking consciousness. These entities are the divine archetypes or Ideas, the self-intelligent intelligibles most full explained by Plato and Plotinus among ancient authors. It is into these that our consciousness reverts in the phase of sleep. Timeless and nowhere, we disappear into a myriad centres, each aware of itself and linked by radiant light to all. We could not know of their being by any kind of knowing we know of here; we may infer it from the hypothesis that mind and world are analogous in the ways I have described, that the stars have their counterparts in our thoughts.

By virtue of knowing themselves the Ideas are self sustaining and therefore eternal. They are the origins of things in this world which are not self sustaining but require the energies of the Ideas to exist at all. Night is eternal, day is temporal. Dreams are influenced by the Ideas directly since the stars appear with the moon, and also indirectly so far as dreams are memories of a waking life itself determined by the Ideas. The reversion of the mind to the Ideas in sleep is not the contraction of waking consciousness into nothing but its expansion into a multitude of subjects by which its individuality is sundered. Returned once more to the eternal, self sustaining Ideas the mind is renewed and the surplus of their renewing energy is another day of waking consciousness. Then awareness collapsed once more into a point, we move among the world of things. The Ideas are the divine milk of the mind's nourishment, as necessary to waking consciousness as milk to the life of mammals. The Greeks called night Euphrone which also means Well-Mindedness.[1]

* * *

We have now met each of the three Graces, and for a moment let us contemplate just how identical the triplets are. We already see a little way into the occult meanings of their clasping hands, the causal connections between the three realms. There are various problems with the doctrine as I have stated it, unavoidable in a short account. The sun is most erect at midday, so why is dusk the analogue of intercourse? What of the testes and sperm which are central to the act of love and unmentioned? Is not Artemis goddess of childbirth rather than Aphrodite? What exactly is the analogue of the stars in the female body, the womb, the breasts or the milk of the breasts? I have started with a summary account to help the reader grasp the plot from the beginning because it is not a plot in which events follow one another but one in which they are all of them simultaneous. The understanding must advance on many fronts at the same time without actually going much further than it has already been.

The order in which I have presented the three Graces and their realms is not, however, arbitrary. I have started with what is generally known about Venus, that she is a particular planet and the goddess of love, and have gone on from there to what is less generally known, her

---

1. *Hesiod, The Homeric Hymns and Homerica,* op. cit., 'Works and Days', 540.

power over mental events. I have tended to begin from the signs in the heavens when giving the analogues between the three realms because the heavenly signs are clearer and more straightforward than the sexual and mental ones. The reader may feel that while there is some persuasive force in many of the analogies between the three realms, that force often depends on the particular order in which they have been presented; if they had been presented in no particular order they would have been much less persuasive. This is true, I think, only so far as the analogies gain some of their force from their all pointing in the same direction, and this direction I have made clear from the start. It is not, I think, true that certain of the analogies are acceptable only because they derive from other analogies more acceptable in themselves. That certain analogies are more acceptable than others at this stage comes from the fact that we are more used to them, not from intrinsic merit. As analogies they must all be equally valid.

This is a methodological point. The analogies are valid not because they fit a system but each is valid as an analogy in itself. By all being valid in this way they may truly be said to constitute a system. Now it may be that the conviction that there is such a system leads the thinker to discover new analogies. But this is not the same as making them up. This difference between discovering and inventing analogies bears some thinking about. For it shows that what is needed to understand such doctrine as this is not solely the capacity to apply a set of rules and determine their logical consequences, but something else as well, the capacity to recognize an analogy when one sees it. This is a science which requires the heart as much as the head, a sensitivity to certain quickenings of the imagination on first perceiving the inwardness of a connection between two realms. There is an inward sensation of its rightness and though one may have been led to look just there by certain expectations of how the system must work, there is a quite special turning over of the heart when the expectation is met by the discovery. Often there is no such delight, the expectation is not met, and often an analogy forces itself so on the heart that it must be admitted at whatever cost to the system as then understood. And there are many instances of this in this account. In all cases the heart determines. Progress in this path comes from throwing the heart forward and then going to retrieve it wherever it lands.

It will be said that compared to the controlled and duplicable observations on which modern science is based, these intuitions and the analogies which they intuit are too subjective, fragmentary and inexact to constitute a science. This is why poetry is now thought an ornament rather than a system of knowledge. Before we proceed further with the study of ancient works, we must clear the ground of these prejudices. The intuitions of poetic metaphor are not subjective since in the case of at least ancient poetry they have been proven again and again in innumerable hearts over millennia. Certainly it is worth asking whether the chord which a metaphor strikes is strung by nature or convention, but that the chord be struck again and again is the one prerequisite for a metaphor's survival. Systems of poetic metaphor are the first science of a people and shared by them all, while our science which is based on observations made in unusual circumstances can never be a science of the people but only of the few. Unfortunately this scarcely comprehended science of the few is displacing the science of the people in the people's own minds and the old systems of metaphor survive in language and custom rather than in the conscious understanding. Now when glimpsed they appear fragmentary where once they were known whole.

Nor can it be allowed that these metaphors or analogies are inexact. The appreciation of a metaphor is nothing else than the perception of its mathematical exactitude. The sun rising above the horizon, the head lifting from the pillow are analogous just so far as head is to pillow as sun to horizon. This spatial correspondence is typical in the precision of its proportion. The Greeks called it *kosmos*, order or pattern. Analogy is from a Greek word which translated into Latin as *ratio*, a word which we have preserved in this exact sense in English. It is well known that the Greeks were the first mathematicians in the west whose work has survived and that they used a system of proportions, not quantities. Our algebra is such a system, which explains why the Greeks were able to achieve all the results of algebra up to the third power. But the Greeks did not limit their use of proportions to arithmetic and geometry. They discovered the same ratios in poetry, music, biology, astronomy and the plastic arts. What made this possible was their lack of reliance upon quantitative data. Not limited to the measurements of ruler, scales and clock, they were free to discern ratios of every kind between every field. With greater faith than ourselves in the stirrings of their hearts they were led to the appreciation of vast harmonies coordinating all things in heaven and on earth and they called these harmonies by the names of gods.

Greek science explained gods and goddesses as well as the most common experiences of the people and in this regard extended in at least two directions beyond our own. In comparison to theirs it is our science which appears fragmentary. If science by definition is one and all embracing, then the Greeks were nearer the goal than we are who have not developed their work but forgotten it and started somewhere else. Every day the Greeks felt their gods on their pulses and knew how to see their most commonplace experiences as the material evidence of divine powers. This was not blind faith in some abstract, transcendental deity but a keen appreciation of patterns of ratios which were mathematically exact and which corresponded perfectly to the felt differences and similarities between things. Perceiving patterns in the world they experienced, they attempted to embody these patterns in their technology. But since they were not concerned to change a world which they had discovered to be beautiful already, their technology produced what we call works of art. Here in many different ways they reproduced the patterns in a spirit of piety, the formal properties of their creations echoing the ratios of the gods and goddesses whom they celebrated.

Above all else they were at home in their world, of which they made mathematical sense as it was given them. Their science deepened and organized their experience without seeking to change it. If there was a mystery in the creation it was not, as it is for us, a matter of laws undiscovered for millennia and to be found at last by the artifice of experiment. Their mystery was not hidden but laid open to any who chose to meditate upon it. This meditation revealed connections and correspondences hitherto unnoticed by neophytes, yet symbolically repeated in the art and cult around them when they knew how to look. In this way the course of scientific discovery had already been charted and its goal known. The great sages had lived or were living and were not merely to come. To us this seems unadventurous who have found realms unguessed at by the ancients, but let us remember that our new knowledge has not been added to ancient wisdom but has displaced it. Compared to the Greeks we know next to nothing of our everyday world, of the stars as seen by the eye, of the analogies between bodies and minds. We have lost the trick of reading the world mathematically in terms of sense and feeling, and since these are humanly inalienable, we no longer understand what we are and must be as once it was understood. We choose to forgo this in hopes of a science the direct evidence of which is afforded to almost no one. Even these few are not free of this world they know little about.

The world to the ancient Greeks was like a great theatre or stadium in which the gods deployed their effects for the edification of the human spectators who were to meditate upon them. Everywhere there were signs and tokens placed for their seeing, and the art of interpreting these signs was another gift of the gods given at moments symbolized in their rituals. They returned their thanks as to personal deities who wished to help them clarify their understandings. All things were full of gods, not least the human being whose physical and psychic structure was patterned in the clearest possible way after the most perfect ratios, as was fitting for a creature whose task it was to learn them.

Night	Female	Sleep
Stars	Breast milk	Eternal Ideas
Moon	Menstrual Cycle	Dreaming Self
Sunset	Sexual union	Going to sleep
Venus	Mound of Venus	Wand of Hermes
Sunrise	Birth	Waking up
Sun	Phallus	Egoic Self
Day	Male	Being Awake

4

APOLLO CYRNEUS

THE RIDER IN THE CHARIOT is Parmenides but the chariot is the chariot of the sun. The poet does not ride with the god of the sun, he is the god himself. Parmenides describes himself as a sun god so far as the chariot is the sun. It is a large claim he is making and we must ask how he comes to make it. When Socrates was on trial before five hundred Athenian jurymen he insisted in his defence, according to Plato, that he call Apollo at Delphi as a witness on his behalf. This created an uproar in court as Socrates had known it would.[1] But he was not claiming as much as Parmenides. There is no suggestion anywhere in the ancient tradition that Parmenides was unorthodox in his religious beliefs or practices. On the contrary, of the earlier philosophers he has the most immaculate record in his dealings with gods and people. If the chariot is the sun, Parmenides identified himself with the sun in a way regarded as decent in his own time and after.

The little we know of ancient Greek customs shows this identification of a human with a god to be regular and widespread in the practices of cults. It appears that the Greeks formed associations whose members believed themselves the companions of the divinity whom they worshipped. These clubs inducted new members through a rite of initiation and often both worship and initiation took the form of a procession or revel in which a living representative of the divinity was carried along in triumph among the members of the band. Initiation was a ritual new birth. The central figure, the living embodiment of the group spirit, was called the young man

---

1. *Plato, Collected Dialogues*, op. cit., 'Apology', 20E–21B.

or young woman, *kouros* or *kore*. This was the way of addressing the divinity. Zeus, Dionysos, Hermes, Ares and Heracles were all worshipped in this way and so also was Apollo. They were all young men for these purposes. We cannot determine how far the first lines of Parmenides' poem describe an actual practice. But his identification of himself with the sun and his being called young man by Night make it certain that his poem was part of the religious life of his people and not the work of a private man. Epimenides who slept for years in a cave was called the new young man, according to Plutarch, when he went to Athens about 600 BC and averted the plague which threatened that city.[1]

The membership of these bands and the divinity worshipped varied from cult to cult. In one in Crete, where Epimenides came from, an inscription of the third century BC has Justice and Right, Themis, doing the same work they do in Parmenides' poem.[2] In the Herois ceremony at Delphi maidens escorted Semele.[3] In Parmenides' poem maidens escort the rider but among his last fragments is one which has young men on the right and maidens on the left (XVII). This line may refer to cult practice. It is hard to distinguish the initiatory associations from the processions which accompanied victors in athletic or poetic contests. In all cases there was a tendency to regard the central figure as divine. Pindar sings of the athlete victorious in the games by identifying him with the heroes of his clan or people. The athlete represents the hero by his physical prowess and beauty, above all in running. This is the young man we find everywhere from this period, in stone or bronze, standing naked with one leg slightly forward, arms at sides. There is also the bronze charioteer in the same pose but clothed and holding the reins in front. This statue was dedicated to Apollo at Delphi by the Sicilian city of Syracuse in memory of a Syracusan charioteer who had brought glory to his city by winning at the games.

---

1. Plutarch, *Parallel Lives, I, Theseus and Romulus, Lycurgus and Numa, Solon and Publicola*, tr. B. Perrin (London: Loeb Classical Library, 1914), Solon, 12.
2. Harrison, op. cit., p 6.
3. Plutarch, *Moralia*, 'Greek Questions', 12.

Respect for physical beauty and strength, made formal in the rituals of athletic contests and in statuary dedications, appeared also in erotic relationships between older and younger companions of the same sex. From these bonds came the social form of some of the earliest schools of poetry and philosophy: Sappho and her student poetesses, Parmenides and Zeno, Empedocles and Pausanias,[1] Socrates and Charmides.[2] Here the prowess admired was intellectual or moral rather than physical, but we are told that Zeno and Charmides were more than passably attractive in their persons. The young man was one reborn through an initiatory teaching and his beauty was of the spirit. This view of him goes some way to explaining why many of his statues have been found on tombs, regardless of the age of the deceased. But there appears to have been some tension between the physical and spiritual values ascribed to the young man. Xenophanes, who taught Parmenides, criticized the awarding of state honours and pensions to athletes before intellectuals like himself, for it was from the intellect that good laws and full treasuries came.[3] If this was or came to be the attitude of the Eleans, Parmenides need not have been a champion charioteer as well as everything else to be the young man of his poem, the living representative of a god among his people.

We know something of Elean religious practice from ancient sources. We are told by Herodotus that on the advice of a man from Poseidonia the Eleans established a cult of the hero Cyrnus when they founded their city. Parmenides describes himself as god of the sun, and the hero cults were often assimilated into the worship of gods. One example we have seen is Grynus, eponymous hero of Grynea near Phocaea where the temple of Apollo Gryneios was situated. And there was the Spartan cult of Apollo Carneios. It is certainly possible that Cyrnus too was worshipped as a surrogate of Apollo by the Eleans, and the more likely since at the direction of Pythian Apollo the Phocaeans had founded their colony in Cyrnus, Corsica. The Eleans were anxious to comply with the directions of

1. *Ancilla to the Presocratic Philosophers*, op. cit., Empedocles, frag. i.
2. *Plato, Collected Dialogues*, op. cit., 'Charmides'.
3. *Ancilla to the Presocratic Philosophers*, op. cit., Xenophanes, frag. ii.

this god as best they could in their new circumstances. From another point of view the cults were expressive of the group spirit of their members. The founding of a cult to Cyrnus by the city of Elea made Cyrnus the type of the Eleans, the man they should all strive to be. Parmenides made laws for his city, Diogenes tells us. In that act above all others he would be the living representative of the spirit of his people. But we do not know at what stages in his life he wrote his poem and made laws.

The Megarian poet Theognis wrote poems to a Cyrnus at about this time. Here is one of them:

> More accurate than compass, rule and square must that man be on his guard to be, Cyrnus, to whom the priestess of the god gives an oracle from the rich shrine of Pytho. If he were to add anything, he would never find an escape. Nor if he omitted anything could he outstrip his offence in the eyes of the gods.[1]

This poem must be connected to Herodotus' story of the founding of Elea. There the Phocaeans had been told by the Delphic oracle to found a colony in Cyrnus, Corsica as they thought. That colony they had been forced to abandon despite their great courage, and they remained stateless in Rhegium until the man from Poseidonia proposed a solution to the problem of the oracle. The solution was to found a cult of the hero Cyrnus. The Eleans-to-be surely saved Delphi's veracity and authority by taking responsibility for having misinterpreted the oracle's words. And here in this poem Theognis tells someone called Cyrnus to be most exactly scrupulous in his reading of the Delphic oracle's words. This cannot be coincidence. Theognis is speaking of Elea. Like the advice of the man from Poseidonia, Theognis' poem is an attempt to restore the oracle's claim to prescience in the context of a famous case which called that prescience into question.

Here is another of his poems to Cyrnus:

> I have given you wings on which you may rise and fly with ease over the endless sea and over the whole earth. You will be present at all feasts and banquets, on the tongues of many men; and the young

---

1. *The Penguin Book of Greek Verse,* op. cit.,Theognis, 805–810.

men will sing loud and clear of you during their lovely revels, accompanied by their little clear toned flutes. And, when you pass into the hiding places of the dark earth, into the house of Hades filled with lamentation, not even then, not even when dead will you lose your glory, but since your name will be undying among men, you will be famous, Cyrnus, as you circle round the land of Greece and the islands, crossing the unharvested fish-rich sea, not riding on mortal horses. The shining gifts of the violet-crowned Muses will send you on your way, for you will be with all those who care and who will care for songs, as long as there is earth and sun. Yet I do not have even a little respect from you, but you deceive me with your words, as if I were a little child.[1]

On the usual reading these lines are addressed by an older man to his younger boyfriend who is reproached for giving the poet a difficult time after the poet has made his name famous throughout the Greek world for ever. In his first words, on this account, the poet refers to other poems he has written to Cyrnus before this one, and in the collection of poems attributed to Theognis there are many poems to Cyrnus. We may feel on this reading what the poet clearly did not intend us to feel, that young Cyrnus has some just complaint against a lover more concerned with his own achievements than those of his beloved.

But if we take it that Cyrnus is Parmenides, then the poem is less boastful and more specific. The wings which the poet has given to Cyrnus are not now the wings of his own songs but the very wings of song themselves, on which Cyrnus, become a poet with Theognis' help, has arisen to make his own undying reputation as a poet. Theognis compliments Parmenides throughout by imitating and alluding to Parmenides' own poem. No longer horses, now the gifts of the Muses escort him. The palace of Night is the palace of Hades, the piping shriek of the axle the clear toned flutes of the revellers. The Muses are crowned with violets, the crowns of dusk or dawn between the crowns of unmixed fire and night (XII). More than this is the sense of the sun's chariot circling and circling above the world over sea and land. Cyrnus' descent into Hades is qualified, for even when he is dead he will rise again as the sun does. His continuing is

1. *The Penguin Book of Greek Verse*, op. cit., Theognis, 237–254.

enacted by the poet's long central sentence and by his repetitions of land and sea and earth. Cyrnus will matter to all who care for songs while earth and sun exist. And so he has, and not only through Theognis' lines.

If these lines were addressed to Parmenides as Cyrnus, Theognis was claiming that he had taught Parmenides the art of poetry. There is nothing we know of Theognis to contradict this claim. He wrote at some time during the sixth or early fifth century BC and may well have come from Sicilian Megara not far from Elea. His poem to Cyrnus about the oracle shows his concern with Elea. He says in another poem to Cyrnus that he is world famous but does not please all his townsfolk in Megara.[1] Perhaps he did better in Elea. The quality of this poem is sufficient evidence of the skill he would need to have taught Parmenides. And this even though the poem is an imitation after Parmenides. But this, of course, is why he wrote an imitation, to show that he could do what Parmenides had done and thereby justify his own first words. Like all his work the poem is in elegiacs while Parmenides wrote in hexameters. Elegiacs are a foot shorter than hexameters every second line. This may be why in other poems Theognis calls Cyrnus Polypaides, son of Polypaos the many footed.[2]

The last two lines of Theognis' poem are a masterpiece of disingenuousness. After saying that he has taught Cyrnus and showing he could do as well, at the same time as praising his alleged pupil to the skies, he finishes by complaining of how he is completely baffled and deceived by his pupil's words. If Cyrnus is Parmenides we must all agree with Theognis here. Bafflement is exactly what any perusal of Parmenides' poem brings us to. And so Parmenides himself warns us when Night tells the rider to learn the deceptive order of her words concerning the opinions of mortals (VIII). In his last two lines Theognis alludes again to Parmenides' poem, and in complaining he does not understand it, proves how well he does. Theognis is a true teacher who prides himself finally on the fact that his student has surpassed him.

1. *The Penguin Book of Greek Verse*, op. cit., Theognis, 23.
2. Op. cit., Theognis, 25.

And, of course, if Theognis' Cyrnus is Parmenides in this poem and Theognis knew Parmenides' poem, then Theognis' poem to Cyrnus about the Delphic oracle acquires a quite new meaning. Now Theognis is advising Parmenides to be scrupulously careful in his handling of the oracle which he, Parmenides, has personally been given by the Delphic oracle. In this way Theognis' poem about the oracle 'collapses' Parmenides' poem into the Cyrnus oracle. It is the clearest evidence we have of the link between them except perhaps for the lines in which Night calls her own words deceptive (VIII).

* * *

Night's account of being is original in the history of Greek thought both for its method and for its subject. The method is not so purely logical as may first appear, if logic determines the validity only of an argument. Being is something real of itself, to be seen by the rider at the time of his revelation, its properties neither inferred nor deduced but apprehended directly. The direct apprehension of being and its attributes is the seeing of how being is as it is of necessity, and Necessity is a goddess whom Night describes as holding being in the bonds of limit (VIII). She is the personification of logical necessity, as is Fate who does exactly the same work. So also does Justice, so that the attributes of being are at once necessary, fated and just. The bonds of being are of this intelligible kind, though at other levels they are realized in the bounding surface of a sphere and by the outermost limits of the stars (X). The bonds or limits most often discussed by Night are those which restrain being from coming into being or from passing out of it. These bonds limit being in the sense that being, if it is being, cannot do otherwise than remain being. The bonds are logical necessity and this necessity is demonstrated by the argument. Divine bindings are not uncommon in Greek mythology. Prometheus is fettered to a rock of the Caucasus by Hephaestus,[1] and Homer has Hephaestus trap the adulterers Ares and Aphrodite immobile beneath an invisi-

---

1. Aeschylus, *Prometheus Bound*, tr. H.W. Smyth (London: Loeb Classical Library, 1983).

ble golden net.[1] The late author Proclus thought that this net represented a necessity of connection between these two powers which
was at once physical and intelligible.[2] If Proclus was right, Parmenides was original only in that he was the first to demonstrate
intelligible or logical necessity by argument, as well as representing
it by symbol as Homer had done.

Parmenides is also the first in our literary and philosophical tradition to call his principle by the name of being. Being is only one of
its names, however. He refers to the principle indifferently by the
names 'being', 'to be' and also 'is', and by combinations of these. But
they are all parts of the present tense of the verb 'to be', and the
choice of part is obviously determined by the grammatical struc-
tures of Night's argument. Throughout the argument there is a
relentless stress on these parts of the verb 'to be'. They are affirmed
so many times that it is hard to make Night's points when we read
her speech aloud without stressing these words in a way we would
never think to use at any other time. This is unprecedented in the literary remains. How closely these remains are related to each other
we have seen, but nowhere in the work before or contemporary with
Parmenides' do we find any of these words used with this emphasis.
There are no examples from Homer, Hesiod, Xenophanes or Heraclitus. On these grounds we might say that Parmenides founded the
branch of philosophy called ontology, the science of being. This may
be true as it is certainly true that he is the first western exponent of
this science whose work has survived to us. But there is evidence of
very uncertain date that Parmenides' and his followers' stress on the
parts of the verb 'to be' was not unique in Greek thought.

At Delphi there was a representation of the letter E, which was
the same in Greek as in English, in a prominent place at the entrance
to the temple of Apollo.[3] When this letter was put there we do not
know, but Plutarch who was a priest of Apollo at Delphi in the first
century AD wrote a dialogue concerning this letter which concludes

1. *Odyssey*, VIII, 266–366.
2. *Works of Plato*, tr. F. Sydenham and T. Taylor (New York: AMS Press, 1979),
Proclus, 'On the Fables of Homer', I, pp 189–190.
3. Plutarch, *Moralia*, 'On the E at Delphi', p196.

with the thesis that the E is the Greek word meaning 'Thou art'. This is the second person singular of the present tense of the verb 'to be' as 'is', the principle of Night, is the third person singular. According to Plutarch, the mortal entrant into Apollo's temple was enjoined by the letter E to affirm within himself the god's absolute reality by repeating the word it spelled. We find again the stress on this commonest word of the language. Here is the passage from Plutarch:

> I am therefore of the opinion that the significance of the letter is neither a numeral nor a place in a series nor a conjunction nor any of the subordinate parts of speech. No, it is an address and salutation to the god, complete in itself, which by being spoken, brings him who utters it to thoughts of the god's power. For the god addresses each one of us as we approach him here with the words 'Know Thyself, as a form of greeting which certainly is in no wise of less import than 'Welcome'; and we in turn reply to him 'Thou art,' as rendering unto him a form of address which is truthful, free from deception, and the only one befitting him only, the assertion of being.

> The fact is that we really have no part nor parcel in being, but everything of a mortal nature is at some stage between coming into existence and passing away, and presents only a dim and uncertain semblance and appearance of itself; and if you apply the whole force of your mind in your desire to apprehend it, it is like unto the violent grasping of water, which, by squeezing and compression, loses the handful enclosed, as it spurts through the fingers; even so reason, pursuing the exceedingly clear appearance of every one of those things that are susceptible to modification and change, is baffled by the one aspect of its coming into being, and by the other of its passing away; and thus it is unable to apprehend a single thing that is abiding or really is.

> 'It is impossible to step twice in the same river' are the words of Heraclitus, nor is it possible to lay hold twice of any mortal substance in a permanent state; by the suddenness and swiftness of the change in it there 'comes dispersion and at another time, a gathering together'; or, rather, not at another time nor later, but at the same instant it both settles into its place and forsakes its place; 'it is coming and going.'

> Wherefore that which is born of it never attains unto being because of the unceasing and unstaying process of generation, which ever bringing change, produces from the seed an embryo, then a babe, then a child, and in due course a boy, a young man, a mature man, an

elderly man, an old man, causing the first generations and ages to pass away by those which succeed them. But we have a ridiculous fear of one death, we who have already died so many deaths, and still are dying! For not only is it true, as Heraclitus used to say, that the death of heat is birth for steam, and the death of steam is birth for water, but the case is even more clearly to be seen in our own selves: the man in his prime passes away when the old man comes into existence, the young man passes away into the man in his prime, the child into the young man, and the babe into the child. Dead is the man of yesterday, for he is passed into the man of today; and the man of today is dying as he passes into the man of tomorrow. Nobody remains one person, nor is one person; but we become many persons, even as matter is drawn about some one semblance and common mould with imperceptible movement. Else how is it that, if we remain the same persons, we take delight in some things now, whereas earlier we took delight in different things; that we love or hate opposite things, and so too with our admirations and our disapprovals, and that we use other words and feel other emotions and have no longer the same external form, nor the same purposes in mind? For without change it is not reasonable that a person should have different experiences and emotions; and if he changes, he is not the same person; and if he is not the same person, he has no permanent being, but changes his very nature as one personality in him succeeds to another. Our senses, through ignorance of reality, falsely tell us that what appears to be is.

What, then, really is being? It is that which is eternal, without beginning and without end, to which no length of time brings change. For time is something that is in motion, appearing in connection with moving matter, ever flowing, retaining nothing, a receptacle, as it were, of birth and decay, whose familiar 'afterwards' and 'before', 'shall be' and 'has been', when they are uttered, are of themselves a confession of not being. For to speak of that which has not yet occurred in terms of being, or to say of what has already ceased to be, that it is, is silly and absurd. And as for that on which we most rely to support our conception of time, as we utter the words 'it is here', 'it is at hand', and 'now'—all this again reason, entering in, demolishes utterly. For 'now' is crowded out into the future and the past, when we would look upon it as a culmination; for of necessity it suffers division. And if nature when it is measured is subject to the same processes as is the agent that measures it, then there is nothing in nature that has permanence or being, but all things are in the process of creation or destruction according to their relative distribution with respect to time. Wherefore it is irreverent in the case of that which is to say even that it was or shall be; for these are certain deviations, transitions and alterations, belonging to that which by its nature has no permanence in being.

But God is (if there be need to say so), and he is for no fixed time, but for the everlasting ages which are immovable, timeless and undeviating, in which there is no earlier nor later, not future nor past, no older nor younger; but he, being one, has with only one 'now' completely filled 'for ever'; and only when being is after his pattern is it in reality being, not having been nor about to be, nor has it had a beginning nor is it destined to come to an end. Under these conditions, therefore, we ought, as we pay him reverence, to greet him and to address him with the words 'Thou art'; or even, I vow, as did some of the men of old, 'Thou art one.'[1]

This is Plutarch writing as much on the early philosophical tradition as on the E at Delphi. He refers to Heraclitus twice by name, but the account here of the doctrine of change derives also from Socrates' speech in Plato's Symposium.[2] Plutarch like Socrates is concerned with changes in the subject as well as in the objects of perception. Having established the impermanence of time, Plutarch argues that therefore the things in time are all impermanent too. For if the measurer changes, so must what is measured. This is elegant, and so too are the images of the many deaths and grasping the instant like water in the hand. But the greatest influence on Plutarch here, though he is not mentioned by name, is Parmenides, from whom Plutarch has learnt the dialectic of being and the distinction between being and becoming. Parts of Plutarch's account repeat Night's account word for word. What differences there are between Plutarch and Parmenides are matters of emphasis, not substance. Parmenides is showing the connection between being and all seemings, while Plutarch is showing the difference between mortals' opinions of themselves and the god's reality. Plutarch makes clearer what Parmenides had meant by those mortal opinions in which there is no true trust (I). Most instructive of all is the harmony to

1. Plutarch, *On the E at Delphi*, 391–393.
2. *Plato Collected Dialogues*, op. cit., 'Symposium', 207D–208C.

which Plutarch attunes the works of Heraclitus and Parmenides. These two contemporaries, the one the philosopher of change and the other of the utter immobility of being, are presented here as the twin exponents of a single doctrine. Plutarch's synthesis is a corrective to the belief that these early philosophers contradicted each other. They were rather the teachers of a single doctrine; Heraclitus says as much about the divine nature as Parmenides says of what seems, and both are experts in the matter of Justice.

Imagine our standing with Plutarch before the main entrance to the temple of Apollo at Delphi, the sacred center of the Greek world for the past thousand years and more. We are confronted by an inscription which greets each of us with the words 'Know Thyself'. Hanging between the middle columns of the temple is the letter E, as later coins show,[1] and with Plutarch we dutifully reply to the god's greeting with the words 'Thou art'. What all this means on the page we already know and we must now consider what it means in this place. Why does the god greet us here with 'Know Thyself'? We should think of the greeting and response as a kind of examination to establish the respondent's suitability for admission into the temple. The god commands 'Know Thyself' and to comply with the command is to know who you are. But, as Plutarch points out, our identities as mortals are quite unstable. In commanding us to know ourselves the god requires that we know what we are beyond change. But there we as mortals no longer exist, and so the proper reply to the god is that we are, so far as we are, who he is. Our answer to the god's implied question 'Who are you?' is 'You'. For only Apollo can enter the temple of Apollo. This is like a story told by the Persian poet Rumi.[2] A man knocked on a friend's door and his friend inside asked who was there. The man replied 'I am' but his friend inside told him to go away. After a year of wandering the man returned to his friend's house and knocked again. His friend inside asked who was there, but this time the man replied 'You are'. His friend inside said 'Since you are me, come in, O myself!'

1. Plutarch, *Moralia*, op. cit., p196.
2. Jalaluddin Rumi, *The Mathnawi of Jalaluddin Rumi*, ed. R.A. Nicholson (London: E.J.W.Gibb Memorial Trust, 1925–40), 'Mathnawi, I', 3056–3065.

Plutarch's meditation as he enters the temple of Apollo is like Parmenides' poem in more ways than its stress on the verb 'to be'. Orpheus, according to Plutarch, had said that Night and Apollo shared the oracular shrine at Delphi. We must entertain the possibility that Parmenides' entry through the gates of the palace of Night is Plutarch's entry into the temple of Apollo at Delphi. The stone threshold of Night's palace may be the stone threshold of the temple of Apollo in rocky Pytho, to which Achilles refers in the *Iliad*.[1] Inside this temple was the oracle of the Pythian prophetess, a woman like Night, in a dark abode, speaking gnomic truths which were transmitted in hexameters. We have seen how Parmenides, though he speaks in the first person, does so not as himself exactly but as the sun god, and perhaps also as Cyrnus, the living representative of a hero who may have been a surrogate of Apollo. All this, if true, would establish his suitability for admission into the temple, as does Plutarch's response to the god's greeting 'Know Thyself'. Plutarch's response 'Thou art' brings the mortal entrant to an awareness of the god's reality, preparing him for the house of the god by making him assimilate himself to the god. The rider in the chariot is already assimilated. The doors of the palace of Night remain closed to the uninitiated but they open for him after the maidens have persuaded Justice with gentle words. The rider says nothing and the goddess within greets him with 'Welcome' and not 'Know Thyself'. 'Welcome' is also a greeting mentioned by Plutarch in this context. Then she tells him what Plutarch later meditated upon as he entered the temple. But when Night speaks to the rider she is not a mortal entrant addressing the god; she speaks as the temple's divine resident. She does not say 'Thou art' because she is, or rather neither of them but only being is, beyond all discrimination.

Which came first, Parmenides' poem or the E at Delphi? Socrates in an early dialogue of Plato refers to the 'Know thyself' as the god's greeting at Delphi.[2] So that inscription must have been there in the fifth century BC, into which Parmenides lived. Poems sometimes describe or explain ritual practices but they rarely originate them.

---

1. *Iliad*, IX, 404.
2. Plato, *Charmides*, op. cit., 164a–165b.

So if this poem and this practice were related, the E was probably first. What is clear is that Plutarch read the letter in accordance with Parmenides' account of being and placed his reading in the context of a practice very like Parmenides' entry into the palace of Night.

5

DELPHI

BEING IS LIKE the mass of a well rounded sphere stretching equally in all directions from the center (VIII). Night discovers to the rider the unshaken heart of well rounded truth (I). The sphere is the sphere of the outermost stars, of the womb and of sleep, but it has its own center, different from the sun and its analogues which we have supposed the centers against Night's sphere till now. Night speaks from the heart of well rounded truth, her permanent station to which the rider goes and from which he comes. When he is there with Night he sees how being is like a sphere which stretches equally in all directions from the center. The rider's arrival where Night resides and her revelation are intercourse and conception, and also the experience of being awake and asleep at the same time. The mingling in this place of the male and female seeds (XVIII) is analogous to the mingling of the waking and sleeping minds. As the mingled seed expands and grows from the moment of conception, so does the wisdom of the initiate from the moment of his revelation. Sleep too is a place far within, to which we withdraw from the outer world and our physical senses and which contains the whole world. These are psychological and biological analogues to the center of a sphere where Night is, but we have yet to find its correlate in the astronomical realm. Metaphysically, being is no more like a sphere than a point because being is incomparable and so neither big nor small. The sphere and its heart are equally images of being because the sphere is the least discriminated of shapes. It is the principle of spatial extension as being is the principle of everything.

Where is the center of the starry sphere if it is not the sun? What in the macrocosm corresponds to the heart of thought and of the womb? It is our earth, which we now call a planet but which is unshaken in our immediate experience. Earth is also the stone threshold of the gates of the paths of Night and Day, over which the rider in the chariot passes. The gates are surrounded by a lintel, the surrounding heaven (X). Stretching from earth to heaven, the gates join the dark center of Night with the distant periphery of the fixed stars. Earth is mentioned twice in the later fragments and she was one of the most ancient divinities worshipped at Delphi. [1] To her, first of all, the oracle there was said to have been sacred. To think of the earth as at the center of Night is to take the immediate appearance of the sky with the seriousness which Parmenides always accords appearances.

We have now to ask what corresponds in the astronomical realm to the mingling of the male and female seeds in conception and to the union of the waking and sleeping minds in revelation. Until now we have supposed dusk to be the astronomical phenomenon which corresponds to sexual intercourse and going to sleep. But we are not now concerned with these but with conception and revelation, not with what happens in the sky as seen from the earth but with the earth itself. We have, as it were, come in from the sky when we crossed the stone threshold and we are now at the center inside the palace of Night where conception and revelation take place. In Hesiod's story of the birth of Aphrodite, the sky god Uranus constantly impregnates the earth goddess Gaea until Cronus castrates him and makes the sun set. The impregnation of the earth by the sky is the play of sunbeams on the earth when the sun illuminates the darkness at the heart of Night and causes the germination of earthly creatures. Here is the innermost shrine where light and night are made one. The rare moment of conception, the rarer moment of revelation are analogues to the light of common day on earth. Apollo's entry into his temple at Delphi was the coming of summer.

Our earth is at the center of the starry sphere. But the mingling of male with female, waking with sleeping or sun with earth at the

---

1. Aeschylus, ii, *Agamemnon, Libation-Bearers, Eumenides, Fragments*, tr. H. Weir Smyth and H. Lloyd-Jones, (London, Loeb Classical Library, 1983), 'Eumenides', 2.

heart of things is the concentration of contraries into a single point. This point is beyond contraries. So this center is neither male nor female, waking nor sleeping, sun nor earth, but the place where each of these actually coincides with its other. This common structure in biological reproduction and the heavens suggests a cosmical genesis corresponding to the animal. The universe too may be thought to be a single, living animal and to begin from a single point which is beyond discrimination into contraries, now because it is by definition prior to all else in manifestation. This point brings the universe into existence by stretching out equally in all directions, up and down, left and right, forward and back. To these six directions we must add the seventh, the center itself through which alone there can be a return to the unmanifested principle. The point at the center which is the first of all manifestations is closest of all to the unmanifested. This idea is very clearly expressed in the following account from the Kabbala:

> After recalling that the Holy One, blessed be He, the Unknowable, can be apprehended only through His attributes by which He has created the worlds, let us begin with the exegesis of the first word of the Thorah: Bereshith (In the beginning). Ancient writers have informed us regarding this mystery that it is hidden in the Supreme Degree, the pure and impalpable Ether. This degree is the sum total of all the later mirrors. They proceed from it by the mystery of the point which is itself a hidden degree emanating from the mystery of the pure mysterious Ether.
>
> The first Degree, absolutely concealed, cannot be apprehended. Similarly, the mystery of the supreme point, though profoundly hidden, can be apprehended in the mystery of the inward Palace. The mystery of the supreme Crown corresponds to that of the pure and impalpable Ether. It is the cause of all causes and the origin of all origins. It is in this mystery, the invisible origin of all things, that the hidden point, whence all proceeds, takes birth. On that account, it is said in the Sepher Ietsirah: 'Before One, what canst thou count?' Which means: before that point, what canst thou count or comprehend? Before that point, there was nothing except Ain, that is, the mystery of the pure and impalpable Ether, so named (by a simple negation) by reason of its incomprehensibility. The comprehensible beginning of existence lies in the mystery of the supreme point. And since this

point is the beginning of all things, it is called Thought. The mystery of creative Thought corresponds to the hidden point. In the inward Palace the mystery attached to the hidden point can he understood, for the pure and impalpable Ether remains for every mysterious. The point is Ether rendered palpable (by the concentration which is the starting point of all differentiation) in the mystery of the inward Palace or Holy of Holies.

Everything, without exception, was at first conceived in Thought. And if anyone should say: 'Lo! there is something new in the world,' impose silence on him, for that thing was previously conceived in Thought.

From the hidden point emanates the inward Holy Palace (by the lines issuing from that point along the six directions of space). This is the Holy of Holies, the fiftieth year, which is likewise called the Voice that emanates from Thought. All being and all causes thus emanate by the power of the point from on High. Behold that which relates to the mysteries of the three supreme Sephiroth.[1]

This passage brings together a number of the elements in Parmenides' poem and in the same way. We may identify the Crown or supreme principle with unmanifested being, being in its metaphysical aspect; the hidden point is the unshaken heart of well rounded truth, the center from which being expands equally in all directions; the inward palace is the palace of Night. More similar still is the way in which the passage brings the three worlds of thought, space and genesis into a single order, identical with the union of the psychological, astronomical and biological realms according to Parmenides. The passage enables us to appreciate what is involved in Parmenides' doctrine concerning the center of being. This must be quite clear in principle if we are to understand it in its application. The doctrine of the sphere of being explains the place and cult of Delphi at the center of the ancient Greek world.

Inside the temple of Apollo at Delphi was a certain stone, the navel stone or *omphalos*, which was supposed by the Greeks to mark the center of the inhabited world as the navel does of the human body. The story went that two eagles or swans flying in opposite

---

1. René Guénon, *Symbolism of the Cross*, tr. Moses de Leon (of excerpt) and A. MacNab (Hillsdale, NY: Sophia Perennis, 2003), pp27–28.

directions from the furthest parts of the earth had met in Delphi at this stone. These birds may yet be seen perched on the legends of maps made today. Plutarch tells of two priests also, the one coming from Britain and the other from beyond the Persian Gulf, who met at Delphi.[1] The navel stone was hemispherical and sacred to the goddess Earth. It was not the center of the earth but of the earth's surface, or of that part of its surface inhabited by man. This was why it was called the navel, which is the corresponding center to the surface of the belly.

The stone was inside the temple of Apollo. Also inside the temple was the oracular shrine of the Pythian prophetess. As in a womb, *delphys*, male and female, waking and sleeping, light and night were joined here. What was done here enacted the hidden point of the cosmic genesis and organized the ancient Greek world after its pattern. The *omphalos* showed that this was where the womb was. From within, the hidden point expanded as the priestess directed her enquirers to found colonies at the ends of the earth. This was the womb in which these cities were first conceived in thought, a sanction to make them permanent like the world itself.

Parmenides' poem helps explain what this center meant and why the greatest oracle of the Greek world was there. The hidden point of the cosmic genesis which was represented at Delphi is also the heart of thought, in which all things lie in their principle. The center of space is also the center of thought. The poem explains the name of the stone marking the center. It was called the navel stone because it marked the representation of the cosmic womb. In this place the heart of space, the heart of thought and the heart of generation were at one. It is usual for oracular shrines to be placed at geographic centres. In the *Vedas* it is said that all oracles are set as is the nave within the wheel.[2] A word for an oracular pronouncement in Greek is *omphe*, which is connected with *omphalos*. From here the Greek world opened out, guided by the principial thought at its center.

Seven directions were primary, represented by the six points of a three dimensional cross and its center. These seven directions were

1. Plutarch, *Moralia*, 'On The Obsolescence of Oracles', 409E–410B.
2. *The Hymns of the Rg Veda*, ed. T. H. Griffith, 1963, Benares, *Rg Veda*, VIII.41.6.

realized in the cult and doctrine of Delphi. The upward direction on
the vertical axis was implied in the doctrine that Apollo went from
Delphi as swift as thought to join the other gods on Olympus. We
are told this in an early hymn to Pythian Apollo, probably composed
before the time of Parmenides.[1] The downward direction was represented by the chasm beneath the oracular shrine from which prophetic vapors rose, according to Strabo, and over which the priestess
was seated on her tripod when she gave her oracles.[2] Tartarus,
according to Hesiod, was as far beneath the earth as heaven was
above it.[3] The four lateral directions, north, south, east and west
were usually represented by winds. In the Hymns of Orpheus, of
uncertain date, there are five hymns grouped together, to Aurora, to
the north, south and west winds, and to the goddess Themis.[4]
Aurora, dawn, is east. Themis, Right, is the goddess whom Night
names with Justice as having brought the rider to the palace. Right
was a great power at Delphi, according to the Orphic hymn, with
authority over the oracle. The grouping of the hymns indicates that
Right was at the center of the four quarters. This was the seventh
direction, the hidden point at the center of the six, which is first in
manifestation and the way to the unmanifested. This direction was
symbolized by the Delphic temple and by the proper response to the
god's greeting there.

Justice too is at the palace of Night. From the doors of the palace
she pushes out the bar so that they may open. But like Night Justice
is within and without. Night is at the center, here on earth, and at the
periphery of the starry sphere. Her gates stretch from the one to the
other. Justice is within the palace and also holds the sphere of being
in the limits of great bonds, as though she were the power constrain-
ing its surface. We might think of her as exercising this power from
the center, except that Justice is Aphrodite, goddess of the star of east
and west. Aphrodite was worshipped at Delphi, the center of the

---

1. *Hesiod*, op. cit., 'Hymn to Pythian Apollo', 186–187.
2. Strabo, IX, 3.5.
3. *Hesiod*, op. cit., 'Theogony', 720.
4. *Thomas Taylor the Platonist*, op. cit., 'Hymns of Orpheus', LXXVII–LXXXI,
pp 285–287.

sphere, but Night places Justice in the midst of the crowns of fire and night and this means her place as Evening and Morning Star.

We cannot tell from what we have of the poem whether Parmenides thought this star nearer to the earth than were the limits of Heaven. But even if Justice does not herself mark the limits of heaven in the east and west she governs these directions in another way, by governing the movements of the sun in relation to them. In the same way we must suppose her to govern the directions north and south. In Greece in summer the sun's circuit is far north of the equator, in winter far south. This alters the length of the day and the times of the sun's rising and setting. If Justice controls the course of the sun she must therefore control two different axes, from east to west and from pole to pole. In this way she holds being in great bonds, like Right among the three winds and Aurora in the Hymns of Orpheus. The star also crosses the meridian at zenith and nadir.

Delphi was the location of Mount Parnassus, a favorite place of Apollo and the Muses. Here, we may suppose, the Muses inspired Parmenides, the poet of Delphi, as they had inspired Hesiod generations before on the slopes of Mount Helicon. Here is another meaning for the maidens who escort his chariot. Theognis sings of the gifts of the Muses which accompany Cyrnus. Mount Parnassus was the Mount Ararat of Greece where Deucalion, the Greek Noah, first made landfall after the flood. The Roman poet Ovid describes the receding of the waters from this point to reveal the land all round.[1] This ebbing also symbolizes the emergence of a world from the primordial point of creation.

* * *

Night concludes what she has to say about truth and being and then tells the rider to learn about mortal opinions (VIII). According to Night mortals make the mistake of supposing that there are two separate things in the world. Their mistake is that they do not realize that these two must be one if they are to be understood. Finally only being is comprehensible and being is one; whatever else is understood

---

1. Ovid, *Metamorphoses*, tr. F. J. Miller, rev. by G. P. Gould (London: Loeb Classical Library, 1977), 315–347.

must be understood in its relation to being from which all things exist. Any plurality must be resolved into this unity. But this resolution escapes mortals who name two distinct forms and think of them as quite separate from each other. Night says that these two separate forms are fire and night. They think of fire as gentle and nimble and of night as unknowing, dense and heavy. Night herself says this which shows how wrong mortals are to think night unknowing.

Who are the mortals who posit the existence of fire and night and think of them as separate from each other? Night has no particular group in mind but supposes that we all make this mistake, that the making of this mistake is a condition of mortal life. She represents the error in terms of fire and night because the absolute discrimination of the one into some two is the simplest and clearest form of the error, and because fire and night are among the subjects of the poem's opening lines. There the chariot of fire penetrates the palace of Night to enable that unity of thought which is the understanding of being. Having shown how mortals distinguish between fire and night, the goddess shows how these two are one in the later fragments of the poem as they are in the first. She declares that to think of anything as either light or night is mistaken because everything is equally full of light and night. There is nothing whatever which does not share in both light and night (IX). This is false only if we grant the absolute distinction between light and night in the first place. But this is the distinction Night rejects.

Then Night tells the rider he is to learn about the ether and all the works of sun and moon, how heaven came to be and how Necessity bound it to hold the limits of the stars (X, XI). Where is the rider to learn these things? It may be that Night refers here in the future tense to later parts of the poem which have not survived. The only surviving parts of the poem which account explicitly for the generation of the universe are the line that Love was the very first of all the gods to be devised (XIII) and the lines about the world's birth (XI). These are not adequate as a cosmogony by themselves. It may be that Night is not here referring to parts of the poem yet to come, but is telling the rider what he will learn from other sources than the poem. This is unlikely in a poem so closely woven together. A third possibility is that Night is referring in the future to what the rider will learn from his study of her words in time to come. In this case there is a cosmogony in the poem but it will appear to the rider only after long contemplation (II). The cosmogony is realized implicitly in the poem and the rider will have to work hard to find it. This is the most likely explanation. There is a cosmology to be derived from the earlier and later parts of the poem, but it takes time. If this is how we take Night's reference to the future, we may compare her words to the rider with those of Empedocles to his pupil Pausanias, urging him to press what he has heard deep into his mind and study it, because he will obtain many other things from it.[1] These early scientists were not concerned to give a straightforward description of things, but such an account as would lead their students to understand the principles from which any description must be derived.

But there remain many things which Night claims the rider will learn and which we have not. What we have lost of the poem matters here. Our interpretation of Parmenides has given no place to the moon but Night does (X, XI, XIV, XV). Nor have we yet a full picture of the sun's movements. There is the binding of heaven by Necessity to hold the limits of the stars (X). This may refer to the sphericity of the heavens, and there is a passage in Plato's Republic, similar to these lines of Parmenides, which suggests a further meaning. Socrates is describing the journey of souls about to be born back into the world:

> Now when the spirits which were in the meadow had tarried seven days, on the eighth they were obliged to proceed on their journey, and, on the fourth day after, he said that they came to a place where they could see from above a line of light, straight as a column, extending right through the whole heaven and through the earth, in color resembling the rainbow, only brighter and purer; another day's journey brought them to the place, and there, in the midst of the light, they saw the ends of the chains of heaven let down from above: for this light is the belt of heaven, and holds together the circle of the universe, like the under-girders of a trireme. From these ends is extended the spindle of Necessity, on which all the revolutions turn.[2]

1. *Ancilla to the Presocratic Philosophers,* op. cit., Empedocles, frag. cx.
2. Plato, *The Republic,* op. cit., 616B.

Necessity and the limits of heaven, but here the limits are more elaborately described, like the girdles of triremes holding the world together. The Greek trireme was strengthened by the passing of ropes right round its hull to hold it in. These limits sound, in fact, like lines of longitude or latitude such as we use ourselves to measure the surfaces of spheres.

There is a tradition that Parmenides was the first to say that the earth was round,[1] but this discovery was also attributed to his predecessors Hesiod and Pythagoras, and Strabo thought that Homer knew it.[2] There is a stronger tradition that Parmenides was the first to divide the surface of the earth into five zones, probably two polar, two inhabited and a burnt one in the middle. Poseidonius, according to Strabo, said that Parmenides made the burnt zone almost twice as big as the tropic zone, extending well into the temperate zones on both sides.[3] Aetius tells us that Parmenides was concerned to define the inhabited regions of the earth in relation to the tropics.[4] A man of early Elea would have had to have been a practical geographer after his own experience and those of his Phocaean ancestors. About 300 BC, Pytheas, a citizen of Phocaean Massalia and the first Greek known to have circumnavigated Britain, established the latitudinal bearing of Massalia with a gnomon at the summer solstice.[5] The words of Delphic Night about the limits of the stars reflect an interest in such things on Delphi's part two centuries earlier. Night makes her speech to a possible surrogate of Apollo, god of mathematics and geography.

There is another passage in Plato's *Republic* which, though famous for its obscurity, may yet be brought to bear very directly on Parmenides' poem and on these words of Night. At this point in the dialogue Socrates is explaining how the ideal political constitution which he and his companions have discussed will, since it is created, at some time be dissolved. This dissolution begins at the top when the philosopher rulers cease to breed as they should:

1. Diogenes Laertius, op. cit., VIII, 48.
2. Strabo, I, 1.20.
3. Strabo, I, 9.4.
4. Diels & Krantz, op. cit., Aetius, III, 11, p226.
5. Strabo, II, 5.8.

How, then, Glaucon, I said, will disturbance arise in our city, and how will our helpers and rulers fall out and be at odds with one another and themselves? Shall we, like Homer, invoke the Muses to tell 'how faction first fell upon them,' and say that these goddesses playing with us and teasing us as if we were children address us in lofty, mock serious tragic style?

How? Somewhat in this fashion. Hard in truth it is for a state thus constituted to he shaken and disturbed, but since for everything that has come into being destruction is appointed, not even such a fabric as this will abide for all time, but it shall surely he dissolved, and this is the manner of its dissolution. Not only for plants that grow from the earth but also for animals that live upon it there is a cycle of bearing and barrenness for soul and body as often as the revolutions of their orbs come full circle, in brief courses for the short-lived and oppositely for the opposite. But the laws of prosperous birth or infertility for your race, the people you have bred to be your rulers will not for all their wisdom ascertain by reasoning combined with observation but they will escape them, and there will be a time when they will beget children out of season. Now for divine begettings there is a period comprehended by a perfect number, and for human by the first in which augmentations dominating and dominated when they have attained to three distances and four limits of the assimilating and the dissimilating, the waxing and the waning, render all things conversable and commensurable with one another. Whereof the basal four thirds wedded to the pempad yields two harmonies at the third augmentation, the one the product of equal factors taken one hundred times, the other of the same length one way but oblong, a hundred of the numbers from the rational diameters of the pempad lacking one in each case, or from the irrational lacking two, and a hundred cubes of the triad. And this entire geometric number is determinative of this thing, of better and inferior births. And when your guardians, missing this, bring together brides and bridegrooms unseasonably, the offspring will not he wellborn or fortunate.[1]

The obvious difficulty with this passage is the number which Socrates propounds. This number is the key to successful marriages and it is often called the nuptial number. Surprisingly, Socrates' complicated number is less difficult to make out than it looks, and there is general agreement among scholars that the number which

1. *Plato, The Collected Dialogues,* op. cit., *Republic,* 545D–546D.

Socrates has in mind is 12,960,000 or, as we might say, sixty to the fourth power.[1]

The 'number for human begettings' is 36, which is the smallest number with four sets of two factors, the 'four limits', of which three sets are of unequals, the 'three distances'. For 36 = 18 x 2; 12 x 3; 9 x 4; and 6 x 6 (6 x 6 is not a distance [lit. 'apostasy'] because its factors are equal, but it is a limit). These four sets of factors may be represented as four plane rectangles. In each of these, the longer side, the larger number, is the base. Of the numbers 18, 12, 9, and 6, only 12 is four-thirds of another whole number. This 'basal four thirds' multiplied by 5, the pempad, yields 60, which is multiplied by itself three times.

For our purpose we need examine only the last of what Socrates tells us about his number, that it yields two harmonies. The first of these harmonies, we are told, is the product of taking equal factors one hundred times. Let us divide 12,960,000 by 100 and find the square root of the remainder. The square root of 129,600 is 360. So our first harmony is 360 x 360 x 100 = 12,960,000. The second harmony is more complicated. The first factor is common to both harmonies, 100. The second factor is one of the numbers from the rational diameters of a pentad each lacking one, or from the irrational diameters each lacking two. The third factor is one hundred cubes of the triad. The triad is 3; 3 x 3 x 3 = 27; 100 x 27 = 2,700. So we have the first and last of the three factors of our second harmony, 100 and 2,700. Let us divide 12,960,000 by both of these factors. The result is 48. This must be the number from the rational diameters of the pentad each lacking one, or from the irrational diameters each lacking two. And so it is generally agreed to be.

We have our two harmonies, whatever they are. The first is 360 x 360 x 100, and the second is 100 x 2,700 x 48. Why are we given only these two sets of factors for the nuptial number? 12,960,000 is divisible into many sets of three factors. Why does each of our two sets have three factors rather than two or four or any other number? Let us suppose that what Socrates has in mind are two rectangular

---

1. James Adam, *The Republic of Plato* (Cambridge: Cambridge University Press, 1902), II, pp204–208, 264–312.

solids. Cubes are sometimes called harmonies because they have six sides, eight corners and twelve edges, and 6:8:12 is an harmonic progression. Rectangular solids have the same property. These solids are measured by their three dimensions of length, breadth, and height. The multiplication of these three dimensions by each other gives the volume of the solid. We know also that our two sets of three factors have one factor in common, 100. Let us make 100 the height of each of our solids. Then we may imagine the first solid as 360 long, 360 broad and 100 high and the second solid as 2,700 long, 48 broad and 100 high. The first solid is square and squat, a cube cut off more than two thirds of the way down. The second solid is oblong, very long and thin. But they are the same height and have the same volume.

Two of the six faces of the square, squat solid are squares, 360 x 360, and these two squares are its base and its top. The other four faces of the square, squat solid are rectangles, 360 x 100, and these four rectangles are its sides. All six faces of the long, thin solid are rectangular. The two rectangles at its ends are 48 x 100, the two rectangles at its base and top are 48 x 2,700; the two rectangles on its long sides are 100 x 2,700. Imagine these two solids resting on the ground in front of us. Since this is the nuptial number, let us drive the long, thin solid like a spear through the square, squat solid. We drive one of the two smaller sides of the long, thin solid (48 x 100) through the middle of one of the sides (360 x 100) of the square, squat solid. Our geometrical manipulation is meant to represent a marriage and it is where male and female join which most attracts our eyes when we witness these things in the flesh. We may represent the conjunction of the solids by this plane figure:

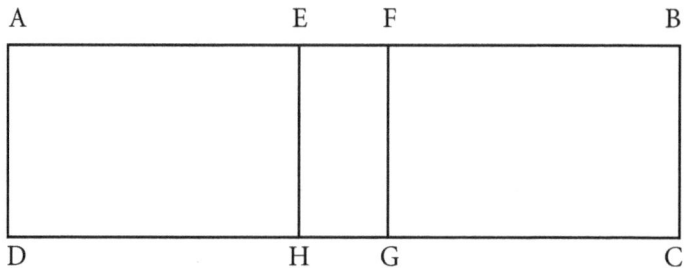

ABCD is one of the four sides of the square, squat solid; EFGH the area cut out of the middle of ABCD by the passing through it at right angles of one of the two smallest sides of the long, thin solid. AB and DC are 360. AD, EH, FG, and BC are 100. EF and HG are 48. AE, DH, FB, and GC are equal.

AB and DC are 360, and 360 is the number of the degrees at the center and on the circumference of a circle. Let us take it that Socrates intended AB and DC as rectilinear representations of the circumferences of two circles. EF and HG cut out an arc of 48 degrees each from these two circles. If, standing somewhere on the surface of the earth, we draw a line on the earth through the two poles and through where we stand, we make a great circle of 360 degrees, the center of which coincides with the center of the earth. This great circle is called the meridian of a place by astronomers. If now we mark out on this great circle the extreme points north and south which the sun reaches directly overhead in its annual course and we then measure the distance between these two points, we find that it comes to nearly 48 of our 360 degrees. This is the distance in degrees between the tropics of Cancer and Capricorn. Later Greek authors computed the distance between the tropics and equator as equal to the angle subtended at the center of a regular fifteen sided polygon by each of its sides. This angle is 24 degrees.

Let us now imagine another great circle, this time drawn through the celestial and not the terrestrial poles and passing directly over where we stand. This circle, though much larger than the one drawn on the earth, we may also measure into 360 degrees. If we now mark out on this circle the extreme points north and south which the sun reaches overhead in its annual course and then measure the distance between these two points, it again comes to nearly 48 degrees. So let us take it that in our figure AB represents the celestial meridian, DC the terrestrial meridian, and EFGH the area cut out between heaven and earth by the annual movement of the sun north and south, so far as we observe it from most places. This last qualification is necessary because the sun actually cuts out 48 degrees of each circle on both sides of the earth, 96 of the 360 degrees of each meridian. The numbers which Socrates gives us represent the sections of arc cut out by the sun on one side of the earth only, in relation to the two complete circles of the celestial and terrestrial meridians.

This interpretation of the nuptial number may be unhistorical in its assumption that Socrates and Plato divided circles into 360 degrees. We know that this division was not generally applied to circles until later. But the division of the zodiac into 360 degrees, each degree a day of the sun's annual course, was certainly older than Plato. The application of this measure to other astronomical circles, in this case the meridians, is a smaller step than its application to all circles and one we might expect to precede it. There are also in the *Republic* those bonds holding the universe together like the girdles of triremes. Socrates attributes the nuptial number to the Muses and it is the number which his philosopher rulers must know to breed properly. The constitution which Socrates has described in the *Republic* up to this point he believes to be the primordial constitution from which all others have devolved. For all these reasons we must take it that Socrates did not present his number as original with himself but of great antiquity.

If we construe the number as measuring the sun's course across the meridians, then our figure of the two rectangles stands very well for the gates of Parmenides' palace of Night. These gates are high as heaven just as the lines EH and FG measure the distance between the celestial and terrestrial meridians. Above the gates is the lintel EF, an unusual lintel which surrounds its doors (I). This is the section of the celestial meridian directly under which the sun passes. The lintel is curved because it represents an arc of the celestial meridian. The stone threshold is HG, the section of the terrestrial meridian between the tropics. We may even draw into the figure an analogue for the two doors of Night's palace. For if we draw a line parallel to EH and FG, of the same length and exactly between them, this line will represent where the doors meet by connecting the celestial and terrestrial equators. The doors open one after the other as the sun moves above and below the equator on its annual course. Since both doors open to Parmenides' chariot, we may perhaps take it that his journey is an equinoctial one. When we put this with Night's words about the bonds holding the limits of the stars and with the tradition that Parmenides was the first to divide the earth's surface into zones, it seems that Parmenides knew Socrates' figure or had a similar one of his own in mind.

On the other hand there are features of Socrates' number which suggest that he was thinking of the words of Night. From his number Socrates generates two sets of three factors which enable the construction of two solids. The first of these we may call night, the second fire. The two solids are then married in the way we have seen. The question must arise whether there is not some simpler way of arriving at the construction which measures the sun's passage across the meridians. But in Parmenides' poem Night tells the rider that mortals make the mistake of naming two forms of which there is no need that there be one, and that they call these forms night and fire. So when Socrates gives the mathematical account of the relation between night and fire, between heaven, earth and sun, he derives his two forms from a single number. He affirms their identity quantitatively at the same time as he distinguishes between their shapes. This is profound but it may also impose limitations on his construction. For once the factors 360, 48, and 100 have been explained, we ask about the other 360 and 2,700, the lengths of the two solids. We may reason that the squares 360 x 360 at the top and base of our first solid both represent two axes at rightangles to each other, the one the meridian north and south as we have seen, and the other the great circle east and west round the equator. But this does not explain 2,700. Perhaps the factors which have not been explained are the result of the constraint on Socrates' procedure that both his solids have the same volume. The number of this common volume is neither night nor fire but the one before these two, symbolizing their unity at the heart of things.

Let us now add detail to the plane figure, shading the inner rectangle where the two solids interpenetrate and marking the line between the celestial and terrestrial equators where the doors meet. What this figure suggests is a very good reason for Socrates to think that his harmonies have something to do with human generation. This is as good a picture of the bride on her wedding night as a rectilinear mathematics is likely to provide. It fairly represents the sun god's view of his way to the heart of desire. The figure now illustrates the similarity between the astronomical and biological formations and we see that on this interpretation Socrates follows Parmenides in the making of analogies between day and night and male and female. The figure goes some way to explaining a confusing tradition that Parmenides related the global zones or directions to animal generation. According to Aetius, Parmenides supposed that the parts toward the arctic regions generated males, the parts toward the south females. Aetius also tells us that Empedocles arranged it the other way round.[1] This is too little to go on, but our figure in its two-fold meaning brings together the meridians north and south and the left and right sides of the womb. Galen supposed Parmenides' line about boys on the right and girls on the left to refer to the parts of the womb (XVII).[2]

The figure also represents the main entrance to Apollo's temple at Delphi and its two doorposts north and south. The early *Hymn to Pythian Apollo* describes how the god came to Krisa, a foothill facing west beneath snowy Parnassus, and there decided to site his temple.[3] Then he laid out all the foundations, wide and very long. The longer sides of the building were aligned east and west; the shorter sides, in one of which was the main entrance, north and south. The temple was not parallel with the equator but inclined toward it in the west. The eastern pediment showed Apollo in his chariot from the front. Guided by Justice Apollo surveyed the earth from his chariot, going from west to east in his yearly journey round the zodiac. In his temple he was the geographer of the Greeks. His laying out of its foundations symbolizes his nature. The lines he drew on the surfaces of earth and heaven ran straight as the arrows he shot from his bow. With this bow he killed Tityus [4] and also the dragon from whose decay Delphi was called Pytho.[5] These were victories of the mind which measures rather than of the reason which overcomes passion. He is pictured on many vases seated on the omphalos inside the temple, his bow and quiver hanging from a peg. The omphalos is covered with a net or knotted ropes, other symbols of geodesy.

This is Apollo Loxias, Apollo the oblique, a title which describes at once the slanting of the sun's yearly course to the equator and the indirection of his oracles. Apollo Loxias was, most of all, the god of

1. Diels & Krantz, op. cit., Aetius, v 7, p 227.
2. Diels & Krantz, op. cit., Galen, *On Epidemics,* VI. 48, p 227.
3. *Hesiod,* op. cit., 'Hymn to Pythian Apollo', 287–299.
4. *Odyssey,* XI, 576–581.
5. *Hesiod,* op. cit., 'Hymn to Pythian Apollo', 357–358.

Delphi. He is the subject of Parmenides' poem and the solution to the Muses' number of generation. According to one late author, the Pythagoreans believed that

> The zodiac runs obliquely on account of the generation of those earthly things which become complements of the universe. For the passage of the sun and the other planets effects the four seasons of the year which determine the growth of plants and the generation of animals. For if these moved evenly, there would be no change of seasons of any kind.[1]

In *On Coming to be and Passing away* Aristotle describes the effects of these heavenly movements in a way which recalls not only Plato's account of the nuptial number but the circular movements of the same and the different in the *Timaeus*:

> It is not, therefore, the primary motion which is the cause of coming to be and passing away, but the motion along the inclined circle; for in this there is both continuity and also double movement, for it is essential, if there is always to be continuous coming to be and passing away, that there should be something always moving, in order that this series of changes may not be broken, and double movement, in order that there may not be only one change occurring. The movement of the whole is the cause of the continuity, and the inclination causes the approach and withdrawal of the moving body; for since the distance is unequal, the movement will be irregular. Therefore, if it generates by approaching and being near, this same body causes destruction by withdrawing and becoming distant, and if by frequently approaching it generates, by frequently withdrawing it destroys; for contraries are the cause of contraries, and natural passing away and coming to be take place in an equal period of time. Therefore the periods, that is the lives, of each kind of living thing have a number and are thereby distinguished; for there is an order for everything, and every life and span is measured by a period, though this is not the same for all, but some are measured by a smaller and some by a greater period; for some the measure is a year, for others a greater or a lesser period.
> The evidence of sense perception clearly agrees with our views; for we see that coming to be occurs when the sun approaches, and passing away when it withdraws, and the two processes take an equal time.[2]

1. *The Pythagorean Sourcebook and Library,* op. cit., Photius, 12.
2. Aristotle, vol. III, *On Coming to be and Passing away,* tr. E.S. Forster (London: Loeb Classical Library, 1978), 336A–336B.

6

ATHENS

IN THE ANCIENT LITERATURE which has survived to us, Greek and Roman, there is only one account of Parmenides in person. Plato in one of his later dialogues, called *Parmenides* or *Concerning Ideas*, describes a discussion between Parmenides, Zeno and Socrates in Athens at about the middle of the fifth century BC. It is doubtful whether this discussion ever took place. Parmenides was born about 540 BC according to Diogenes Laertius. Socrates was born in 469 BC. On this dating Parmenides was over seventy at the time of Socrates' birth. According to Plato, however, Socrates was extremely young at the time of the meeting, Zeno nearly forty and Parmenides very nearly sixty-five.[1] Even if we suppose Socrates fifteen, Diogenes' date and Plato's story are inconsistent. But Diogenes knew Plato's dialogue and must have been aware of the inconsistency. In his *Life of Zeno* Diogenes says that Zeno never visited Athens but spent his whole life in Elea.[2] As for the arguments which Socrates presents at the meeting, and Parmenides' refutations of them, they belong to the later period of Plato's own philosophical development. The dialogue presents Plato's latest thoughts at the time of writing, let us say 370 BC. But the meeting is supposed to have taken place nearly a century earlier, so that what Socrates and Parmenides are said to have said they could not have said then. But if the content of the discussion is Plato's invention, so may the meeting be. Diogenes seems to have believed that it was and Proclus in his

---

1. *Plato Collected Dialogues*, op. cit., 'Parmenides', 127B–127C.
2. Diogenes Laertius, op. cit., IX, 28.

commentary on Plato's *Parmenides* interpreted the circumstances of the meeting symbolically.[1]

On the other side Diogenes' dates also are open to question. He appears to have based them on the assumption that teachers are always forty years older than their students. Xenophanes according to Diogenes flourished about 540 BC, which means that he was born about 580 BC. Parmenides flourished about 500 BC and Zeno about 460 BC. The battle of Alalia, fought about 540 BC by Parmenides' people, was the greatest sea battle in the Mediterranean of that century and Elea was the last Greek colony founded in the west. Battle and foundation were therefore important dates in Greek history. It is possible that Parmenides' birth was believed to have been contemporary with these events because they were well known. There is a dramatic fitness in the notion that Parmenides, the greatest philosopher of his people, should have been born at the time of their crisis. Nonetheless I prefer this earlier date for Parmenides' birth. Diogenes seems very sure of himself despite his knowledge of Plato's dialogue, and I am distrustful of Plato's superlative. He describes Socrates as extremely young at the time of the meeting. This phrase strikes me as the qualification of an author who knows that there is a discrepancy between his narrative and the facts.

The conversation which Plato narrates is a fiction and probably the meeting too. That Plato adopts the device of attributing his own thoughts to his predecessors is characteristic of the Pythagorean tradition to which Parmenides, Socrates and Plato belonged. The later Pythagoreans referred many of their discoveries to Pythagoras himself, no doubt because they believed these discoveries to be implicit in the earliest doctrines of their school. In the same way Plato refers his own thoughts to those masters from whom he had learnt most, to Socrates commonly and here to Parmenides and Zeno. We must not assume that Socrates, Parmenides or Zeno could have said what Plato has them say in his dialogue, but we must also recognize the possibility that Plato believed these attributions of doctrine to be true to the spirit of their teachings.

Among the many later writers who illuminate the doctrines of

---

1. *Thomas Taylor, Works of Plato,* op. cit., Proclus, 'On Plato's Parmenides', III, p 535.

Parmenides, Plato is unique in giving an account of Parmenides in person. How reliable is he here? In the case of his portrait of Socrates we have other contemporary accounts from Xenophon[1] and Aristophanes[2] with which to compare it; in the case of Parmenides Plato's portrait is the only one. We know at least that Athenaeus, a late writer, records a vigorous objection to Plato's description of Zeno as Parmenides' boyfriend in earlier years, though Plato says only that this was what people said.[3] Diogenes Laertius says that Zeno was Parmenides' boyfriend.[4]

The meeting which Plato describes takes place in the house of one Pythodorus outside the walls of Athens. Pythodorus is the narrator of the event in the first instance and we are told in another dialogue that he was a student of Zeno.[5] Parmenides and Zeno have come to Athens to take part in a festival of Athene, the great Panathenaea, and are staying with Pythodorus. Pythodorus describes Parmenides as very grey, Zeno as tall and both of them as handsome. Socrates and others go to the house of Pythodorus to hear Zeno read out his book, newly published in Athens. We are told later that Socrates had visited Pythodorus' guests before. When he and his companions arrive on this occasion, Parmenides and Pythodorus are out somewhere but Zeno is there and reads them his book. He has almost finished when Parmenides and Pythodorus return. They know the book already. Zeno finishes his reading and Socrates asks him to repeat the first hypothesis of the first argument, presumably the very first sentence of the book. Zeno does so and Socrates then summarizes the argument of the whole book in terms which Zeno approves. According to Socrates Zeno's book provided arguments for supposing that there could not be many, for if there were many, impossible consequences would follow.

1. Xenophon, IV, *Memorabilia and Oeconomicus, Symposium and Apology,* tr. E.C. Marchant and O.J. Todd (London: Loeb Classical Library, 1923).

2. Aristophanes, II, *Clouds, Wasps, Peace,* tr. J. Henderson, 'The Clouds' (London, Loeb Classical Library, 1998).

3. Athenaeus, V, tr. C.B. Gulick (London: Loeb Classical Library), XI, 505 ff.

4. Diogenes Laertius, op. cit., IX, 28.

5. *Works of Plato,* op. cit., 'Alcibiades', II, pp 55–56.

Socrates then turns to Parmenides and remarks that Zeno is Parmenides' friend in other things and in his writing too. Zeno, Socrates says, has written his book in the same way as Parmenides had written his poem, but has made some changes to deceive his audience into thinking that he is saying something different from Parmenides. Parmenides had said in his poem that all is one, according to Socrates; Zeno in his book that there are not many. In this way they seem to say something different but are in fact saying the same thing. What is more, Socrates says, they both seem to think that Zeno's book is well over the head of everyone else present.

These remarks of Socrates, even if jocular, have an edge. They suggest that Zeno is a plagiarist and Parmenides and Zeno arrogant. According to Plato, Socrates was half Zeno's age and Parmenides was an elderly man. It is rude of Socrates to address his seniors in this way. But his bad manners seem much worse when we remember that Socrates is only a day visitor to the house of Pythodorus while Parmenides and Zeno are honoured resident guests, a long way from their own city. By talking to Parmenides and Zeno in this way Socrates runs the risk of offending their host Pythodorus as well, whose first duty would be to protect his overseas guests from unpleasantness in his house.

Socrates had addressed his remarks to Parmenides but it is Zeno who answers. Yes, he says gently and proceeds to explain himself. But first he compliments Socrates on his grasp of the book's argument, and thus quietly disposes of Socrates' taunt that he thinks it over everyone's head. He denies intending to deceive people by claiming someone else's work for his own and says that he sees no value in such deception. He explains that the book was written to defend Parmenides against those critics who reduced to absurdity Parmenides' proposition that there is one. The book was to show how the proposition that there are many is equally absurd. Finally Zeno points out that he himself had not published the book, which had been done without his permission. The book, he says, was written in a youthful spirit of controversy rather than from personal ambition.

This is the tactful reply of a gentleman completely unruffled by Socrates' behavior. We may feel that he gives ground a little when he distances himself from the publication of his book. He is not now committed to defending it to the same extent. But what is impressive about his reply is its calmness and accuracy. Zeno takes Socrates' personal attack point by point and answers it as dispassionately as if he were engaged in a geometrical or dialectical exercise. He says he wrote his book in a youthful spirit of controversy and surely, we are left to understand, Socrates can sympathize with that.

I accept all this is as you say, Socrates says and returns to the attack. Zeno had written that if there are many, these must be both like and unlike, which is impossible. So there cannot be many. Socrates now suggests that this argument is trivial. Of course things are at once like and unlike, he says, just as they are at once one and many. There is no difficulty in supposing that things have contrary properties when they are considered in different ways. Someone might say that Socrates is many because his left side is different from his right, his upper parts from his lower, his front from his back; he might equally say that Socrates is one, being one man among the seven men present. Socrates asks Zeno whether he does not think that a thing which is like or unlike, one or many, is so by sharing in the idea of likeness or unlikeness, unity or plurality. Some things share in likeness, others in unlikeness and others again in both and there is nothing remarkable about this. What he would like to hear, Socrates says, is an argument which revealed the same confusion in ideas, which showed that likeness itself was unlike or unity plural. He finishes by saying that he is sure Zeno has dealt most courageously with his subject, but it is not a subject in which he, Socrates, takes much interest.

By this time Pythodorus is very uneasy. He describes how he expects Parmenides and Zeno to be angry with Socrates at any moment. But they listen attentively to what Socrates says and exchange smiles of approval with each other. There is in this, I think, something of that exclusive intimacy to which Socrates objected earlier. When Socrates has finished speaking, Parmenides at last enters the conversation and begins by complimenting Socrates on his passion for argument. Like Zeno he seems to want to save Socrates and their host from any embarrassment which Socrates' behavior may have caused. In this regard Zeno and Parmenides have excellent manners.

Socrates had addressed Parmenides when he made his personal attack on Zeno, but it was Zeno who answered. Socrates addressed Zeno in this latest speech but it is Parmenides who answers. Parmenides, it seems, was not prepared to engage in a personal defence of himself and Zeno. Instead Zeno handled Socrates very tactfully and by praising his grasp of the book, led Socrates into propounding his philosophical objections to it. Without in any way registering offence Zeno established the discussion at its proper level, so that Parmenides could enter it with no loss of dignity. Parmenides' smile would show Zeno that he was ready, and Pythodorus that there were no grounds for concern. It is, characteristically, a minimal but most effective gesture, his first in the story. Aphrodite is often called a lover of smiles.

Parmenides begins by complimenting Socrates on his passion for argument and goes on to demolish him. He now does for Zeno exactly what Zeno had done for him. In his book Zeno had defended Parmenides from his critics by showing how the arguments of those critics led to consequences as absurd as the ones they derived from Parmenides' poem. But Socrates attacked Zeno's defence of Parmenides for not distinguishing between ideas such as likeness itself and things. Parmenides now attacks Socrates' attack on Zeno's attack on Parmenides' critics by showing how Socrates' theory of ideas, the basis of his attack on Zeno, also leads to insupportable consequences. Zeno defended Parmenides by attacking his critics' position, and Parmenides defends Zeno by attacking Socrates' position.

First Parmenides asks Socrates if he himself distinguishes between ideas such as likeness itself and physical things. Socrates says he does. Parmenides asks him if he believes there are ideas of beauty, justice and good. Socrates says there are. But when Parmenides asks him whether there are ideas of man, water and fire, Socrates is unsure. And when Parmenides asks whether there are, all by themselves, ideas of hair, mud and dirt, Socrates reluctantly says no. Parmenides then points out the inconsistency, telling Socrates he is a young man and that philosophy has not yet possessed him as he, Parmenides, confidently expects that it will. Socrates, Parmenides says, still respects the vulgar opinion that things like hair

270 SCIENCE AND RELIGION IN ARCHAIC GREECE

and dirt are vile, and so cannot admit that there are ideas of them. These remarks of Parmenides are consistent with his poem. There the goddess tells the rider that he is on a road far from the path of people and she frequently distinguishes between truth and mortal opinion. She denies that anything is less than anything else, because everything is full of being (VIII).

Next Parmenides attacks Socrates' claim that ideas are separate from physical things but inform them. Socrates had said that things which are like are like because they share in the idea of likeness itself, but likeness itself is all by itself, which is why it can never be unlike. Parmenides now leads Socrates through a dozen arguments against this separation of ideas from things and Socrates proves unable to sustain his position. For example, Parmenides asks if big things are big by sharing in the idea of bigness. When Socrates says yes, Parmenides asks if bigness itself and other big things are big by sharing in a second idea of bigness which makes the other big things and the first idea of bigness big. For if what is big is big by virtue of a separate idea of bigness, then the idea of bigness can only be big by virtue of a second idea of bigness and so on indefinitely. Or Parmenides asks Socrates whether the knowledge of ideas is the same as the knowledge of things. Ideas are separate from things, according to Socrates, so the knowledge of ideas cannot be knowledge of things. But the knowledge of ideas, as Socrates agrees, is more accurate than the knowledge of things. The gods presumably have this better knowledge, but then they cannot know things by means of it. So we who know things do not thereby know ideas, and the gods who know ideas do not thereby know things. So we know nothing of the gods and they know nothing of us, a conclusion which Socrates can neither avert nor approve.

Most readers of Plato's dialogues have found Socrates' separation of ideas from things as hard to accept as Parmenides finds it. In this dialogue Plato has Socrates acknowledge the problem, which Parmenides expounds. Parmenides had said in his poem that being is one, that being is close to being and that nothing breaks its continuity (VIII). The goddess also told the rider that the things which seem had to be, all of them passing through everything. Nothing is separate from them. The goddess knows very well the circumstances of the rider's journey, but if the gods knew nothing of our world there could be no such oracles as hers.

Socrates cannot defend his position against Parmenides' arguments but he does not run away. His resilience is exemplary. He shows he can take hard knocks in debate as well as give them, and that he deserves Parmenides' compliment on his passion for argument. The compliment would have strengthened his resolve but it is a long and hard test. Parmenides finishes his attack by telling Socrates that there are many other difficulties with his theory of ideas beside the ones already raised. And then he relents, agreeing with Socrates that there must indeed be ideas which remain the same, however hard it may be to prove it to others. For if there were no such ideas, he says, nothing could be said which would remain true. Socrates gratefully assents.

It is apparent now that Parmenides was not criticizing Socrates for believing in ideas but for confining their range and for separating them from things. Parmenides too believes that there are ideas according to Plato. What is there in Parmenides' poem to justify this attribution to him of a belief in ideas of all kinds and inseparable from things? There is nothing about ideas in his poem. But there are first of all the horses, gates, chariot, palace and so on of the opening lines. The variety of their meanings makes these things more universal than their earthly counterparts of the same names. But they are not called ideas, only things which seem, all passing through everything. Yet they had also to be, the goddess says and so confounds in advance, as it were, Plato's customary distinction between beings and seemings, ideas and things. If, on the basis of what we have of the poem, Plato attributed to Parmenides a belief in ideas of all kinds and inseparable from things, Plato must have supposed that Parmenides had thought of the things which seem as he himself thought of ideas. But the things which seem are beings for Parmenides because they all pass through everything, not because they are separate. So Parmenides' attack on Socrates' separation of ideas from things follows from a reading of his poem according to which Parmenides believed in ideas. More clearly, the goddess is concerned with ideas in her proofs of the nature of being. She establishes dialectically the relations between being and unity, plurality, generation, motion, rest and so on. The goddess does not call these terms ideas either, but the relations between them are perceived by the reason and not the senses (VII).

Socrates is now in a quandary. Ideas are necessary to discourse, Parmenides has said, but his own theory of ideas is in ruins. Where will you turn now for your philosophy, Parmenides asks him and Socrates admits that he cannot quite see his way at present. So Parmenides tells him what he has been doing wrong. He has been too forward in trying to define beauty, justice and good. He should first exercise himself much more thoroughly in what the world considers idle nonsense while he is yet young, if he wishes to reach the truth. What do you mean, Socrates asks. What Zeno was doing in his book, Parmenides replies and fully repays Socrates for his earlier rudeness. But immediately Parmenides congratulates Socrates on having wanted to apply Zeno's method to the ideas of discourse as well as to things. This softens Socrates' fall. Parmenides explains to him how Zeno had considered not only what followed from the proposition that there is one, but also what followed from the proposition that there are many. With any hypothesis, Parmenides says, one should examine what follows both if it is true and if it is not true, in regard both to what is hypothesized and to everything else, in their relations to themselves and to each other. This is the procedure to follow with likeness and unlikeness, one, many, motion, rest, generation, corruption, being and not being and any other idea one may choose. If Socrates had done this with his ideas, he is given to understand, he would not now be in the difficulties he is.

Parmenides' list of ideas repeats many of the terms of the goddess in her account of being. And the double method with hypotheses is hers in the poem too. She explores the two hypotheses that it is and that it is not, and by following through the consequences of both she produces her various proofs about being. For example, she argues that being is not generated on the grounds that since being is, it is not in time and so not generated. But she argues the other way when she asks from what being was sprung if it was generated. The answer cannot be not being since there is no such thing to be thought or said, and therefore being is not generated. This follows from the absurdity of supposing that it is not.

Socrates does not have problems because he believes that there are ideas, but because he does not understand them. Parmenides understands them, Socrates now realizes, and Zeno too. We are not given any indication of what Socrates felt when Zeno's book which he had openly disdained turned out to be a model to him on how to correct his own weaknesses. It must have been the more painful to discover this about the book after the pleasure he had taken in Zeno's compliment on his grasp of it earlier. Parmenides just makes the point clearly. He is far from exacting any revenge. But we see another reason now for Pythodorus' uneasiness before. Socrates had been lecturing his elders on what they understood very much better than he did.

But Socrates is not completely discountenanced even now. When Parmenides has finished giving his programme of dialectical exercises, Socrates tells him that it sounds like an enormous undertaking and that he is not sure he fully understands it. But why didn't you go through all this for me and give an example, he asks Parmenides. Socrates had told Zeno at the beginning of their discussion that he wanted to hear how ideas were as complex and confused as Zeno had shown things to be. Parmenides had answered him then, but not simply to give him what he wanted. Instead Parmenides had attacked his theory of ideas. Parmenides must have felt that Socrates needed to be brought to a quandary first, if he were to hear what he had asked for to any good end. Socrates could not learn anything while he thought he knew it all. But this is not the answer Parmenides gives him now. He tells Socrates merely that he is asking too much of a man his age. So Socrates turns to Zeno and asks him to go through it. It was Zeno whom Socrates had asked in the first place and I suspect that Parmenides has just used his age to excuse himself so that Socrates should ask Zeno next and Zeno be caught out. In any case Socrates' asking of Zeno repairs all the damage done by his earlier rudeness, since in asking Zeno Socrates shows he assumes that Zeno could have given the account of ideas for which he had asked.

But Zeno laughs and avoids answering Socrates by joining Socrates in asking Parmenides. It is a great task Parmenides has spoken of, Zeno says, and if there were more of us present it would not be proper to ask him. But as it is I join Socrates in his request, Parmenides, and would like to hear you teach again after all these years. So says Zeno, and then the other four join him and Socrates in asking Parmenides to demonstrate what he means. So Parmenides is caught out and replies:

> I cannot refuse, although I feel like the old racehorse in Ibycus who trembles at the start of the chariot race, knowing from long experience what is in store for him. The poet compares his own reluctance on finding himself, so late in life, forced into the lists of love, and my memories too make me frightened of setting out, at my age, to traverse so vast and hazardous a sea of discourse. However, I must do as you wish, for after all we are what Zeno says we are. Where shall we begin, then? What supposition shall we start with? [1]

The poem of Ibycus to which Parmenides refers has been preserved:

> Yet again will Love eye me tenderly from beneath dark brows and cast me with manifold magic into the endless nets of Aphrodite. I swear his approach makes me tremble like an old champion horse of the chariot race when he draws the swift car all unwillingly to the contest.[2]

Ibycus was a native of Rhegium on the toe of Italy, the Greek city to which the Alalians escaped after the sea battle and from which they founded Elea. It is proper that Parmenides refer to the most famous poet of his people's benefactors. Ibycus was contemporary with Xenophanes, and though little of his work survives we can see that some of its subjects were the same as Parmenides'. Ibycus spoke of the pillars which support heaven as slender,[3] of a silver egg from which twins were sprung,[4] and of Orpheus whom he called Orphes.[5] The slender pillars which support heaven sound like the invisible lines drawn by astronomers between heaven and earth. Simplicius, a

1. *Plato. Collected Dialogues*, op. cit., 'Parmenides', 137A.
2. *Lyra Graeca II*, tr. J.M. Edmonds (Cambridge: Harvard University Press, 1924), Ibycus, frag. II.
3. Ibid., frag. XXXI.
4. Ibid., frag. XXXIV.
5. Ibid., frag. X.

late commentator on Parmenides, mentions a silver egg as a poetic fiction of Orpheus and compares it to Parmenides' sphere of being.[1] We may add the name of Ibycus to those of Ameinias, Xenophanes and Theognis on our list of Parmenides' teachers.

Parmenides compares his reluctance to begin the exposition to the reluctance of an old racehorse to race again, to which Ibycus had compared his reluctance as an old man to fall in love again. Having established the image of the chariot race Parmenides compares the parts of his exposition, as he expounds it, to circuits of the track. It is so long an exposition that by the end of it we admire Parmenides for his gracious consent to run at all. But this introductory speech is more than a courtly acquiescence, since it alludes also to the chariot ride in the opening lines of his own poem. Ibycus and Parmenides have another theme in common, the chariot ride as a metaphor of love. For Ibycus the ride is in a race, for Parmenides the rushing to a palace, but Ibycus' account of the horse's feelings is comparable to the emphasis on the horses in Parmenides' lines. What Parmenides does here and Ibycus does not is to compare the chariot race to the giving of a discourse. In this extension of the metaphor to a third level of meaning we recognize Parmenides as we have come to know him from his poem, in treatment now as well as subjects.

Parmenides, as Plato describes him, and Ibycus, to whom Plato describes Parmenides as referring, both have a tendency to the mixed metaphor. In Ibycus' poem we move from the endless nets of Aphrodite to the racehorse, in Parmenides' speech from the racehorse to crossing a vast sea of discourse. I do not find such shiftings in Parmenides' poem, but in the dialogue he speaks in prose, not verse. 'Sea of discourse' is a fine phrase for the speech Parmenides has just agreed to give. It turns out to be two thirds of the whole dialogue. There is a suggestion that it is hard to ask an old man to make such a speech who has already sailed from Elea to Athens to gratify his friends. Parmenides' metaphor of the sea is happy, too, in the context of his people's history. In these few words we catch an echo of that great journey from Phocaea to Alalia which we know

---

1. Diels & Krantz, op. cit., Simplicius, 'Physics', 146, p 221.

from Herodotus. In Parmenides' poem there is the reference to those whose path was backward turning, an allusion to those who did not undertake the journey despite their oaths (VI). The same man who agrees to undertake the journey in Plato's dialogue criticized those who did not in his poem.

The exposition by Parmenides which follows is twice as long as the dialogue to this point. From here to the very end of the dialogue is one uninterrupted sequence of questions and answers as Parmenides takes the youngest person present, a young man called Aristotle, through the dialectical exercise. Aristotle has little more to say than yes or no, it is Parmenides who does all the work. So Plato's dialogue like Parmenides' poem is largely a single exposition, by Parmenides to Aristotle and the others as by the goddess to Parmenides. In the latter part of the dialogue there are two speakers, Parmenides and Aristotle, but in the poem only the goddess speaks. But Night is addressing the rider throughout, giving commands and asking questions, so that in the poem too there is a communion. In this way Plato's dialogue follows Parmenides' poem structurally as well as topically. In both there is a long concluding exposition and an introduction which brings the reader to the place from which he can understand it. In the poem it is the chariot ride and in the dialogue the discussion with Socrates. But that discussion is at a much lower temperature than the poem. This is not the temple of Apollo at Delphi but a house in Athens; not Night who speaks but Parmenides, and his speech is a dialectical exercise. The goddess reveals the sacred heart of truth. This is the difference between the archaic and the classical, naive and naturalistic art. Plato's description of Socrates, Zeno and Parmenides is lofty as well as lively, but it is not beautiful compared to the eternal moment when Night holds the rider's hand.

Nonetheless this dialogue is superior to all the accounts of the ideas which Plato had written before it and supersedes them. For in those accounts the theory of ideas is expounded as Socrates expounds it here and it is not denied. In this dialogue that theory of ideas is demolished. Parmenides confirms Socrates in his belief that there are ideas, but shows that they are not separate from things and not confined in their range. Socrates had correctly transferred his attention from things in the world to ideas and was to spend his life helping others do the same. But in this dialogue we are shown that this is only the first step and that there is another as great beyond. The dialogue is an account of how Socrates himself came to take this second step and points beyond the distinction between invisible and visible, formal and material. This does not remove the need for the earlier dialogues. Plato's programme requires that we first detach ourselves from the physical world by attending to the ideas as other than the physical world. In this way we accustom ourselves to using the mind's eye and only after this is it possible for us to integrate our experience. To this integration this dialogue points, but it points also to how much more study is needed in the dialectic before we shall hear Night's words concerning being.

Parmenides compares himself to the racehorse to which Ibycus had compared himself falling in love. Is Parmenides' allusion to Ibycus' poem confined to the image of the racehorse or does it extend to the falling in love? Zeno, we are told, was said to have been Parmenides' boyfriend and has just resumed his part as Parmenides' student. Are his the dark brows? Proclus describes Parmenides' exposition as a labour of love which Parmenides approaches with trembling because he must forsake the ineffable truth on which he meditates to go through it.[1] His giving of the exposition is an act of grace, a voluntary remission of his own joy to help others achieve it. But the prime object of this affection is not Zeno who has heard the exposition, nor Aristotle who may not fully understand it, but Socrates who has asked for it. It is Socrates whom Parmenides is engaging with his love talk as elsewhere Socrates engages others. Plato's Parmenides is very like Plato's Socrates. Parmenides' emotion here is Eros and the struggle before which he trembles is the winning over of Socrates. He involves Socrates in his own thoughts to achieve an intellectual union as in the poem Night joins the rider to her own divinity. Justice who is Aphrodite brought that about and Eros appears in Parmenides' reference to Ibycus as Socrates is initiated into the Eleatic dialectic.

---

1. *Thomas Taylor the Platonist*, op. cit., Proclus, 'On Plato's Parmenides', III, p101.

* * *

In his work on rhetoric a later Aristotle mentions that the citizens of
Elea once asked Xenophanes whether they should sacrifice to Leucothea and institute mourning rites.[1] Xenophanes advised that if
they thought Leucothea a goddess they should not mourn; if a mortal, they should not sacrifice. Leucothea was the goddess who saved
Odysseus from shipwreck in Homer's *Odyssey*.[2] Odysseus was over-
whelmed by a storm sent by Poseidon. The goddess rose up from
the bottom of the sea and stood on his raft. She told him to take off
his clothes, tie her veil round himself, leave the raft and start swim-
ming. Odysseus finally did so and reached the magical kingdom of
Phaeacia, from which he was returned to his home in Ithaca. Leucothea was once the mortal lady Ino who was driven mad by Hera,
queen of the gods, and jumped off a cliff into the sea. But Dionysus,
or Aphrodite according to Ovid,[3] took pity on her and made her a
goddess. In the *Odyssey* she saved Odysseus, and the citizens of Elea
may have believed she had done or would do the same for them.
Xenophanes was connected with Elea in its first years, so it may be
that this proposal of a rite to Leucothea was an attempt by the city
to commemorate the circumstances of its founding. In this case the
city wished to give thanks to an appropriate goddess and find an
occasion to mourn their drowned and murdered kinsfolk. But
Xenophanes said no. This way was not open to them since it confused the divine with the mortal.

The cult of Cyrnus had been proposed before the foundation of
Elea by the man from Poseidonia. It was through this cult, I believe,
that the Eleans were given their memorial. A man of Elea, one of the
first to grow up there, wrote a poem which described his journey to
the house of a goddess who revealed the truth to him. This poem
alluded in its opening and elsewhere to the Eleans' experience, but
the goddess who brought the journey to its successful conclusion
was Justice, not Leucothea. In his characterization of Justice the

1. Aristotle, vol. xxii, *The Art of Rhetoric*, tr. J.H. Freese (London: Loeb Classical Library, 1975), 1400B5.
2. *Odyssey*, v, 333–353.
3. Ovid, *Metamorphoses*, iv, 535.

poet combined the suffering and salvation of his people. On her first appearance by name in the poem the poet called her much punishing Justice and later told how she presided over painful birth (XIII). Justice controlled the bar and keys of the gates through which the rider had to pass, but she did let him pass at the maidens' urging. The poet attributed the painful birth of a kouros to the awful will of this goddess, as the events of the *Iliad* were in fulfilment of the will of Zeus.[1] Justice was Aphrodite, the star of the sea, savior of sailors and of Leucothea herself on one late account.

There is nothing unusual about this. If it were all, Parmenides would be a poet of local importance only. But his poem is also, I have suggested, the best and earliest account we have of the doctrines which underlay the practice of the Delphic oracle. The philosophy which Night expounded in the poem was to become the summit of Plato's system. Why was it given to a citizen of the new and distant city of Elea to propagate these truths of central importance? Parmenides was unusually gifted, but we have seen that his work was not so much original with himself as part of a living, sacred tradition. Parmenides had first to be taught what his poem conveyed, which raises the question of why he was taught it and then allowed to transmit it.

The answer may be that it was politic at this time for Delphi to bind Elea to itself. At the end of the sixth century BC. Elea's sister colony Massalia was building a great treasure house at Delphi. Among the Greeks of Asia minor the people who founded Elea were the only ones to do what Bias, the wise man of Priene, later advised all the Ionians to do, leave Asia Minor to the Persians and found cities in the west.[2] This advice of Bias, a man closely connected with Delphi, was in accord with the Delphic policies of western expansion and of retreat before the Persians. The choice of Parmenides to convey central doctrines rewarded Elea for its adherence to these policies by ensuring the city's place in the history of Greek thought. It would be a fit reward for a people who had risked everything to live freely as Greeks.

1. *Iliad,* 1, 5.
2. Herodotus, 1, 170.

Elea was also the city of those who had defeated the Carthaginians and Etruscans in the great sea battle. The Eleans had done much to further Greek arms in the west, against enemies twice their number. There were the ancestral connections between Elea and Phocaea, Phocaea and Phocis, Phocis and Delphi, and the role of Phocaea in the opening of the western Mediterranean to Greek trade. It is possible that the ancestral links between Phocaea and Delphi assisted the geography of this undertaking. Finally there was the story which Aristotle records, that the Eleans themselves asked for a new religious dispensation soon after the founding of their city.

Each of these considerations could have played a part in the making of the poem, but none of them is as suggestive as the story Herodotus tells us about the founding of Elea. The great journey from east to west, the sea battle, the murder of the Alalian sailors and its consequences were epic events. The hexameters of Xenophanes and Parmenides were no more august than the people for whom they were written. The long suffering, courage and salvation of these people were an apotheosis, which brought them close to annihilation and to the gods and which was accompanied by divine portents. In this story, as often in Herodotus and always in Homer, gods and mortals engaged directly; the historical events became legends as they happened. The supernatural consequences of the sailors' murder marked the point at which human history was made sacred.

All this had happened to the Eleans before Parmenides wrote and published his poem. The subject of his poem was another apotheosis, his own, and his poem is one of the fullest and most direct accounts of a man's meeting with the divine to have survived from this very early period. The poem was not for the poet alone but for his people, not about him alone but about his people. The poem was the culmination of their adventures, the crown which they had won.

* * *

The battle of Alalia has been fought and won; the Phocaeans have quit Corsica; the dead sailors are celebrated heroes; Elea has been founded. But Delphi itself had authorized, perhaps even ordered, the founding of Alalia in 565 BC, a sanction to make the colony permanent. Clearly the Cyrnus oracle had failed. From this point of view the exceptional courage and prowess of the Alalian sailors actually compounded the problem: the Alalians could hardly be blamed for the colony's loss.

Herodotus tried to blame them even so: the flood of migrants to Alalia from Phocaea in 545 BC brought about a locally aggressive policy in Alalia; in just five years the Alalians had driven the Etruscans and Carthaginians into a defensive alliance against them. This is implausible. Alalia commanded a fleet of sixty warships and maintained with Massalia the Phocaean thalassocracy in the west. Just as the Phocaeans had had to fight a sea battle with the Carthaginians at the founding of Massalia, so now they had to defend themselves when Alalia was enlarged by the refugees from Harpagus. They would most probably have had to fight a battle of Alalia however they had acted towards their neighbors. This was no reason for the oracle's failure.

There was an earlier explanation: the Phocaeans had misinterpreted the oracle in 565 BC, when they decided to found Alalia. The Greek of the Cyrnus oracle must have allowed a reading different from the Phocaean, according to which a cult, not an island, was the subject of the oracle. This was the suggestion of the unnamed Poseidonian and the subtext of Theognis' poem to Cyrnus about the oracle. The misinterpretation may have turned on very little: only the gender of the definite article distinguishes between the island Corsica and the hero Cyrnus and the definite article is often omitted. But if the Cyrnus oracle was about a cult and not a colony, that cult was surely to be established by the Phocaeans. It had nothing to do with the displaced Alalians who had, on this interpretation, never been authorized as colonists at all. If, on the other hand, the reinterpreted oracle referred to both a cult and a colony, this would have been unprecedented: directions to new colonists that they found a cult but no indication of where to settle. And here the cult hero may have given his name to an island fit for colonization. Certainly the heroes Grynos from near Phocaea and Carnos from Sparta make a cult of Cyrnus plausible in itself, but the theory that the oracle had been misinterpreted was not very plausible at all.

The failure of the Cyrnus oracle was one context of Parmenides' poem. There is an echo of Alalian suffering when Night tells the rider that much-punishing Justice and Right predestined him to take his solitary way, no evil fate. It is a nice balance this, between the harshness of Justice and her justice. All, Night tells him, has always been well: he is still under the jurisdiction of divine and benevolent powers, however severe, however hard to believe. Parmenides' Justice is as grim as the Justice of Heraclitus to whom the Furies minister, or as the Aphrodite of Homer in her dealings with Helen. But she has brought him and admitted him to the ultimate goal of desire.

Could this one man's revelation outweigh all the sufferings of the Alalians? A similar question is sometimes asked about the redemptive power of Christ's passion. For the Eleans the precedent must have been Odysseus, whose journey destroyed all his comrades but also made him, and him alone, an immortal. He could not have achieved immortality but through his dreadful journey. Without its terrors and losses he would have returned as just one more ageing veteran of the Trojan campaign. And so for Parmenides. Without the sufferings of his people, the misdirecting oracle, there would have been no meeting with Night, no knowing the secrets of the Gods. Their losses marked these men out, broke them free of the rest of humanity, and conduced finally to the very greatest good. It was a terrible price they paid for wisdom, these Greeks.

Odysseus and Parmenides gained wisdom at the cost of many other people's lives. This was the price paid by the Eleans and by the Ithacans before them. Almost as compensation, one survivor met God face to face, then told his people of it. From those tellings came a religion in the one case, and in the other a school of philosophy which has long outlasted the religion. In both cases a communal disaster served as a kind of mass human sacrifice, as the catalyst for making one man immortal. Athene stroked Odysseus with her hand on the beach of Ithaca;[1] in the palace Night took the rider's right hand in her hand. The goddesses joined these solitary mortals to themselves and comforted them. At Night's touch the rider sees as God.

1. *Odyssey*, XIII, 288.

PINDAR AT DELPHI

By Olympian Zeus I pray:
O Delphi, golden, glorious
For your oracles! O you Graces
With Aphrodite! Receive me here
In this holy time as prophet
To the singing Muses.

                    Beside
The waters of Castalia
Whose gates are bronze, I heard afar
A murmuring of dance and song
In which no male voice joined. I came
To put this lack of men to flight
Both for the people of your town
And my own honour's sake. Obeying
My dear heart as when a child obeys
His loving mother, down I came
To Apollo's grove, home of feasts
And laurel crowns.

                By the shadowy
Navel of the world, girls of Delphi
Were singing and dancing to the Son
Of Leto, drumming on the ground
With their feet to the nimble rhythm.

                —from *Paean VI*

APPENDIX

THE RECEPTION OF HOMER AS A PHILOSOPHER

Is HOMER'S POETRY philosophical? The easy answer to this question is yes and no. Yes, Homer's poems were generally considered philosophical up to about 1900; and no, Homer's poems have not been considered philosophical for the last century. But though this answer is easy and true, it is unsatisfactory. Homer's poems have remained the same for two and a half millennia: why should they have ceased to be philosophical at the end of this long period? It is not the poems which have changed but our notion of philosophy. Our late exclusion of Homer from the philosophical register tells us much more about us than it tells us about Homer. It tells us that our present view of philosophy and of the history of philosophy is eccentric.

We may begin with that account of Homer and Hesiod given by Herodotus at the end of the fifth century BC:

> But it was only—if I may so put it—the day before yesterday that the Greeks came to know the origin and form of the various gods, and whether or not all of them had always existed; for Homer and Hesiod, the poets who composed our theogonies and described the gods for us, giving them all their appropriate titles, offices, and powers, lived, as I believe, not more than four hundred years ago.[1]

Homer and Hesiod established Greek theology and religion according to Herodotus. Since for every line of Hesiod we have twenty or more lines of Homer, we may take it that the bulk of this theology was Homer's. We may feel that the achievement of this task does not make Homer a philosopher exactly, a lover of wisdom who seeks for it, but one who is already wise. We may compare Homer to Moses, the author of the Pentateuch, as a founder of religion, a

---

1. Herodotus, II, 53.

prophet and patriarch, a figure of scriptural authority. But unlike Moses Homer was both venerated and deprecated by his successors, followed and damned. A century before Herodotus, the philosopher Xenophanes had blasted Homer for his impious treatment of the very gods whose worship Herodotus believed Homer to have established. According to Xenophanes, Homer's accounts of these gods are absurdly anthropomorphic and morally wicked, since Homer describes the gods as engaging in theft, adultery, deception and lying.[1] Xenophanes' disapproval of Homer is echoed a generation later by the crosspatch philosopher Heraclitus, who declared that Homer deserved a beating.[2] But both Xenophanes and Heraclitus would agree with Herodotus that Homer did indeed establish, with Hesiod, the forms of Greek religion.

This simultaneous veneration and deprecation of Homer in classical Greece has no parallel in Judaism or Christianity. Perhaps as a result we find it very hard to accord Homer the kind of spiritual pre-eminence we grant Moses in the history of his people. This is to overlook a difference between the Greeks and ourselves in our approach to religion: classical Greek civilization was essentially dialectical. No other people in our epoch has been as much given to argument as the Greeks. The early attacks on Homer by Xenophanes and Heraclitus turn Greek religion into a matter of the mind. No creeds here, no commandments, no principles. Every step had to be taken by every seeker. This dialectical engagement with their scriptures, with no holds barred, is a defining characteristic of the early Greeks. Xenophanes and Heraclitus, unlike Homer, are still accounted philosophers in our time, and they are attacking Homer as a philosopher.

The dialectical tension between Homeric scripture and the unfettered seeker is screwed to a much higher pitch by Plato's critique of Homer in the *Republic*. Plato attacks Homer on many fronts, building in part on the earlier criticism of Xenophanes. For Plato too, Homer's stories of the gods as adulterers and deceivers are inadmissible and Plato instances the passages in the *Iliad* and *Odyssey* to

---

1. *Ancilla to the Presocratic Philosophers*, op. cit., Xenophanes, frags. 11–15.
2. Ibid., Heraclitus, frag. 42.

which he takes exception. He mentions, for example, Homer's tale of how Ares and Aphrodite commit adultery, and their entrapment by Hephaestus, Aphrodite's husband, under a golden net. According to Plato such stories, even if allegorical, are not suitable for young ears, since children are in no position to distinguish the allegorical from the literal. Further, Plato argues that all poetry such as Homer's which imitates a variety of speakers is deceptive, while a plain narrative of a discussion in indirect speech is acceptable.[1] Plato makes this attack in a literary dialogue in which he carefully imitates in direct speech the contributions of his several speakers!

At the very end of the *Republic* Plato gives us a different view of Homer. In his mythical account of the afterlife with which the *Republic* concludes, Plato's speaker, Socrates, defines philosophy as the knowledge one needs as a discarnate soul when one comes to choose one's next animal or human life on earth. As Plato describes the process by which each soul chooses its next life on earth, he emphasizes that the free choice we make then is the most important one of all. It is for this moment that all our training on earth should prepare us. And then Socrates describes the choices of next lives made by the great heroes of the past, by Agamemnon and Ajax and Odysseus, the heroes of Homer's poems. Odysseus is the last to choose, takes the longest time to do it and finally chooses the most philosophical life, a life most like Socrates' own. Homer's Odysseus stands before us at the very end of Plato's *Republic* as the last paradigm of philosophy. Plato, it seems, could discern in Homer's Odysseus the lineaments of a philosopher.

In the last centuries before Christ and in the first centuries following, Homer's achievements as a philosopher were recognized in a different way, as scientific and symbological. The school of the astronomer Apollodorus read Homer's account of Odysseus' journey as a description of the terrestrial globe, with its arctic, temperate and equatorial zones. This mode of interpretation was developed by Crates of Mallus, the famous librarian of Pergamum, the second library of Greek learning after Alexandria. In the third century AD, the philosopher Porphyry, disciple of Plotinus and

1. Plato, *The Republic,* op. cit., x.

critic of Christianity, developed this way of reading Homer still further in his essay *Concerning the Cave of the Nymphs*. The cave of the Nymphs is the cave which Homer describes on Ithaca as Odysseus' point of return after his nineteen years' absence. Porphyry's essay on Homer's cave of the Nymphs is the single substantial account we have of how to read Homer symbolically. For Porphyry Homer's cave is a symbol of the world and of the womb. Odysseus' arrival there is a symbol of how the immortal spirit is incarnated into a mortal life.

On Porphyry's reading, Homer supposes that the human being is a spirit invested by a body, that we are incarnated. Porphyry interprets Homer as a proto-Pythagorean or a proto-Platonist, despite Plato's attacks on Homer in the *Republic*. This is certainly to see Homer's poetry as philosophical. And we remember that Plato did not exclude the possibility of reading Homer allegorically. He argued only that young minds were incapable of the allegorical reading and should not be given certain tales. Porphyry concludes his essay with the following remarks:

> Nor is it proper to believe that interpretations of this kind are forced, and are nothing more than the conjectures of ingenious men: but when we consider the great wisdom of antiquity, and how much Homer excelled in prudence and in every kind of virtue, we ought not to doubt but that he has secretly represented the images of divine things under the concealments of fable.[1]

Porphyry's conversion of Homer into an honorary Platonist continues with Proclus in the fifth century AD. Proclus wrote an elaborate defence of Homer against Plato's criticisms in the *Republic*. Proclus, too, is seeking to reconcile Homer and Plato, the two greatest teachers of his tradition, and he takes very seriously Homer's claims that he is describing gods in his fables. For Proclus each appearance of the gods in Homer's epics is a theophany, a manifestation of an eternal principle. So for example, Proclus defends Homer's story of the adultery of Ares and Aphrodite and their entrapment by Hephaestus as an account of the divine principles of

---

1. *Thomas Taylor the Platonist*, op. cit., Porphyry, 'Concerning the Cave of the Nymphs', 36.

Love and Hate, how they are bound together by the creator in each and every living creature and will remain bound until the creature's dissolution.[1] Proclus' defence of Homer against Plato is the second most extended work we have of Homeric criticism after Porphyry's essay on the cave of the Nymphs and it appears just before the end of classical paganism. Like Porphyry's essay it is a Platonist work.

In 529 AD, Plato's Academy in Athens was closed by order of the Church after nine hundred years and despite the massive contribution already made by Platonism to Christian theology. The Christian God, who is also the God of the Old Testament, is a jealous God at least in public. In the Eastern Church the study of Homer was maintained but at their best the interpretations were moral rather than intellectual or spiritual. The allegory where it exists has become one-dimensional. Almost exactly one thousand years after Proclus' defence of Homer, the philosophical reading of Homer reappears in Western Europe with a spate of new disciples. Among them is the glorious figure of Pico, Count of Mirandola in northern Italy. This young man picked up the threads of the philosophical Homer exactly where Proclus had laid them down. His writings are full of Proclus and Plotinus, Plato and Porphyry. He knew Latin, Greek, Hebrew, Chaldee and Arabic besides Italian and French. He was trained in Scholastic philosophy and was expert in logic, mathematics and physics. He maintained the transcendental unity of all religions as a kind of universal, holy magic, and he wrote of Homer:

> [We] shall prove someday in our Poetical Theology, that Homer disguised this wisdom, as all other wisdoms, under the wanderings of his Ulysses.[2]

That day never came and we do not have Pico's proof of the philosophical dimension to Homer's *Odyssey*, Pico died at the age of thirty-one.

He died in Florence and it is easy to see why he should have chosen to go there. This was the city where Marsilio Ficino had founded

1. Ibid., Proclus, 'An Apology for the Fables of Homer', pp 509–511.
2. Pico della Mirandola, 'Oration on the Dignity of Man', in *The Renaissance Philosophy of Man*, eds. E. Cassirer, et al. (Chicago: University of Chicago Press, 1948), p 33.

a new Platonic Academy under the patronage of the Medici, with the intention of diffusing the Platonic doctrines which Ficino took to be the foundation and proof of Christianity. Here, too, Sandro Botticelli was painting his pictures of Venus rising from the sea and with the Graces, and still more wonderful, his version of the adultery of Ares and Aphrodite where a clothed goddess calmly watches over her nude, disarmed and sleeping lover. Here again is the Homeric fable, damned by Xenophanes, half damned by Plato and justified by Proclus, reappearing under Botticelli's hand in a most spiritual form.

The revolution in philosophy led by Ficino and Mirandola was the intellectual vanguard of the Florentine and indeed Italian Renaissance. It must have seemed that suddenly the whole world's wisdom had been made available to them, as well as Christendom's. In every scripture known to them they experienced the poignant shock of recognition. And nowhere more so than with Homer, Pythagoras and the later Platonists. In England Thomas More translated a life of Mirandola and revelled in the new liberty. His *Utopia* and Erasmus' *Praise of Folly* demonstrate the extraordinary rapidity with which the new learning was being diffused and the heady, sometimes even flippant joy which it occasioned. At the beginning of the seventeenth century George Chapman made his translations of the *Iliad* and *Odyssey* into English verse.

Before and after the turn of the nineteenth century England experienced its own full-scale restoration of the Homeric and Platonic philosophy in the figure of a single individual. Thomas Taylor was born of humble, dissenting parents in London and was sent at the age of nine to St Paul's School, famous for its teaching of Greek, Latin and Hebrew. But the young Thomas did not approve of the way in which the dead languages were taught at the school and left in his early teens to study mathematics at home. He had already fallen in love with his wife-to-be and their early marriage precluded any chance of his going to university. So he spent his life as a clerk like millions of Londoners after him, in a bank, in a public office and finally as Assistant Secretary to the Society of Arts. But this clerk had an unusual hobby. He was the first English translator of the whole of Plato's works, the whole of Aristotle's works and of the majority of the works of Plotinus, Porphyry, Proclus, Iamblichus, and the other later Platonists. Meanwhile the great universities slumbered on! It is no surprise that Taylor's writing hand seized up from the double pressures of his work and play.

Taylor was no mere devoted scribe. He wrote powerful introductions to his translations as well as freestanding essays. He appended to his translation of Porphyry's *Concerning the Cave of the Nymphs* his own symbolic and spiritual reading of all Odysseus' adventures in the *Odyssey*. He was a passionate believer in the truth and wisdom of his authors and published this fact far and wide. He was ridiculed for it but he also attracted the best minds of the day. John Flaxman, William Blake and Mary Wollstonecraft were all close friends of Taylor. At a remove from personal friendship, Coleridge called some of Taylor's books 'my darling studies' and Wordsworth's *Intimations Ode* turns on a living Platonism which may be traced back to Taylor. Both Keats and Shelley acknowledged their debts to him. But where Ficino and Mirandola created the mental space of the new Academy within the limitations of Christendom, Taylor and his followers denounced the Enlightenment root and branch. This, as Taylor understood it, was the great threat to the true philosophy, the cloud obscuring Apollo's sun. As with the Renaissance, the reappearance of that sun in the person of Taylor coincided with a new movement in the arts.

The issue between Taylor and the Enlightenment was the meaning of the term philosophy itself. Did it mean what Plato and Pythagoras had meant by it or did it mean what John Locke had redefined it to mean? According to Locke, philosophy was properly a kind of conceptual analysis based on the principle that all our concepts are derived from experience. In the first book of his *Essay Concerning Human Understanding* Locke claims to disprove the traditional doctrine of innate ideas. Philosophy is concerned with no other world than the physical, according to Locke, and empiricism is the only way to know it. Locke nowhere refers to Plato in his attack on innate ideas even though Plato is the classical exponent of the doctrine. Nor does Locke address Plato's arguments for innate ideas. For this reason the Enlightenment does not make any dialectical advance on Platonism though it pretends to do so. It merely denounces the philosophical tradition and calls itself philosophy in contradistinction from it. This was Taylor's view of the Enlightenment. For Taylor, following Plato, the natural sciences were far inferior in purity and power to mathematics, theoretical astronomy, music and dialectic. These sciences, studied independently of nature, were for Plato and Taylor the only path to the wisdom which philosophy is seeking.

For Taylor the restoration of the ancient philosophy was also the only defence against the expanding materialism of his age. The burgeoning trade of the new British Empire seemed to Taylor to have expelled all intellectual life. Rarely can there have been a thinker so much set against the grain of his time, and it is cheering to remember that he and Blake were friends. Again and again in his essays and introductions Taylor rams home the charges and fires his broadsides into the mercantilism everywhere around him. These broadsides are beautifully written, as elegant as they are damning, and Taylor is never drawn by his anger to lose his equanimity. He presides over the many centuries of humanity in the certain knowledge that his wisdom will at length come round again. True reason must eventually triumph over the follies of his own time and of any other.

And so it may. But it is certainly true that in the shorter term Thomas Taylor has lost his battle for the true philosophy far more comprehensively than he could ever have imagined. Locke's notion of philosophy, not Plato's, has triumphed. After Locke's model the history of early Greek philosophy itself has been reconstructed in the twentieth century. In the English speaking universities the study of early Greek philosophy is now devoted to the demonstration of how there the first steps were taken towards the empirical sciences as we now practise them. The early Greek thinkers are no more than ground-breakers on the path to the Enlightenment. Whatever in their writings does not conform to this model becomes primitive, mythical superstition from which their reason had started to save them. The poems of Homer have no place in this canon. Parmenides and Empedocles are still accounted philosophers in the new order, but the poetic form of their work is dismissed as a bizarre, primitive impediment to the clear expression of their physics, biology and logic. In short, the passionate dialectical balance between Homer's poems and the philosophical tradition which followed is conjured out of existence as if it had never been. The syntheses of Homer and Plato achieved by Porphyry and Proclus might never have been written.

For Taylor's hostility to the empiricism of his age is returned in full. Incapable of fighting the traditional philosophy on philosophical grounds, the new empiricism simply defines the old philosophy out. To a Platonist today the term academic is painful when used of the contemporary context, since philosophy and the history of philosophy have largely been excluded from the institutions which claim the name of Plato's school today. As for Homer, I should think that fewer than one in a hundred students who have studied Homer's poetry in the last century in a university have any idea of those dimensions to the poems which the later Platonists opened to our understanding. How could they even begin to enquire since Homer is no longer regarded as a philosopher in the modern sense? By such shifts the history of philosophy is made safe for the intelligentsia of a too mercantile nation. The very institutions created to develop and preserve the ancient philosophy are enlisted in its systematic exclusion. It is some small comfort to remember that the original Academy suffered similar abuse after Plato's death.

Even in the later twentieth century there has been one great university exponent of the ancient philosophy, Kathleen Raine. Like Taylor she has maintained the tradition almost alone in a hostile world. Yeats and Blake seem to have led her to Taylor and Taylor to Proclus and Porphyry. Her book, with George Mills Harper, *Thomas Taylor the Platonist*, I found in our campus library twenty-five years ago. It awakened me to the meaning of Homer. Her two volume *Blake and Tradition* examines Taylor's contribution to Blake's formation and contains a full account of Blake's painting of Homer's cave of the Nymphs in the light of Porphyry's essay. Her studies of Yeats, himself a scholar of Blake, reveal the extent of Yeats' debt to this same millennial tradition from Homer onwards. It is salutary and chastening to read the foremost university critic of his day, F. R. Leavis, on the approach taken by Yeats and Kathleen Raine to Blake:

Let me say bluntly that I am not grateful to Yeats for inaugurating the kind of Blake research of which Miss Kathleen Raine is the recognized high priestess in our time. Blind to Blake's genius, it generates blindness, and perpetuates a cult that, whatever it serves, doesn't serve Blake or humanity. The notion that by a devout study of Blake's symbolism a key can be found that will open to us a supreme esoteric wisdom is absurd: and to emphasize in that spirit the part played in his life's work by Swedenborg, Boehme, Paracelsus, Orphic tradition, Gnosticism, and a 'perennial philosophy' is to deny what makes him important.[1]

So Blake and Yeats, too, disappear from the agenda except for the lines which please Leavis. This is no way to study. But, then, for Leavis the great tradition means George Eliot and D.H. Lawrence and perhaps Dickens, not Homer or Plato, Plotinus or Proclus, Botticelli or Thomas Taylor.

1. F.R. Leavis, 'Justifying One's Valuation of Blake', in *The Critic as Anti-Philosopher*, ed. G. Singh (Atlanta: University of Georgia Press, 1982), p18.

BOOK REFERENCES

Adam, James. *The Republic of Plato*. Cambridge: Cambridge University Press, 1902.

Aeschylus. *Suppliant Maidens, Persians, Prometheus, Seven against Thebes*. Translated by H. Weir Smyth. London: Loeb Classical Library, 1983.

_____. *Agamemnon, Libation-Bearers, Eumenides, Fragments*. Translated by H. Weir Smyth and H. Lloyd-Jones. London: Loeb Classical Library, 1983.

Ancilla to the Presocratic Philosophers. Edited by Kathleen Freeman. Oxford: Basil Blackwell, 1971.

Ante-Nicene Fathers. Edited by Roberts and Donaldson. Kalamazoo, MI: Eerdmans, 1981.

Aristophanes. *Clouds, Wasps, Peace* (vol. 11). Translated by J. Henderson. London: Loeb Classical Library, 1998.

Aristotle (Vol. III). *On Coming-To-Be and Passing Away*. Translated by E. S. Forster, 1978, London, Loeb Classical Library.

_____. (Vol. XXII). *The Art of Rhetoric*. Translated by J. H. Freese. London: Loeb Classical Library, 1975.

Athenaeus, Vol. V. Translated by C. B. Gulick. London: Loeb Classical Library.

Ayer, A. J. *Language, Truth and Logic*. Middlesex: Penguin Books, Ltd., 1936.

Blake, William. *William Blake's Writings*, Vol. I. Edited by G. E. Bentley Jr. Oxford: Clarendon Press, 1978.

Boardman, J. *The Greeks Overseas*. Middlesex: Pelican Books, Penguin Books, Ltd., 1964.

Campbell, Joseph. *Occidental Mythology*. Middlesex: Penguin Books, Ltd., 1976.

Coleridge, Samuel Taylor. *Selected Poetry and Prose of Coleridge*. London: Modern Library, Random House Publishing, 1951.

Dante. *The Convivio*. Translated by Richard Lansing, from 'Digital Dante' website.

della Mirandola, Pico. 'Oration on the Dignity of Man', in *The Renaissance Philosophy of Man*. Edited by E. Casirer, et al. Chicago: University of Chicago Press, 1948.

Diels, H. & Kranz, W. *Die Fragmente der Vorsokratiker*. Dublin/Zurich: Weidmann, 1972.

Diodorus Siculus. *History*. Translated by C.H. Oldfather. London: Loeb Classical Library, 1939.

Diogenes Laertius. *Lives of Eminent Philosophers*. Translated by R.D. Hicks. London: Loeb Classical Library, 1925.

Euripides. *Trajan Women, Iphigenia among the Taurians, Ion, Vol. IV*. Translated by D. Kovacs. London: Loeb Classical Library, 1999.

Greek Elegy and Iambus, Vol. I. Translated by J.M.Edmonds. London: Loeb Classical Library, 1914.

Greek Lyric, I. Translated by D.A. Campbell. London: Loeb Classical Library, 1982.

Guénon, René. *Symbols of Sacred Science*. Hillsdale, NY: Sophia Perennis, 2004.

_____. *Symbolism of the Cross*. Translated by Angus MacNab. Hillsdale, NY: Sophia Perennis, 2004.

Harrison, Jane. *Themis: A Study Of the Social Origins of Greek Religion*. London: Merlin Press, 1963.

Heath, Sir Thomas. *Greek Astronomy*. New York: Dover Publications, 1991.

Hesiod, The Homeric Hymns and Homerica. Translated by H.G. Evelyn-White. London: Loeb Classical Library, 1914.

Hesiod. *Theogony, Vol. I, Works and Days, Testimonia*. Translated by G.W. Most. London: Loeb Classical Library, 2007.

Herodotus. *The Histories*. Translated by Aubrey de Sélincourt. Revised by A.R. Bum. Middlesex: Penguin Books, Ltd., 1972.

Homer. *Iliad*. Translated by E.V. Rieu. Middlesex: Penguin Books, Ltd., 1950.

_____. *Odyssey*. Translated by A.T. Murray. London: Loeb Classical Library, 1919.

_____. *Odyssey*. Translated by E.V. Rieu. Middlesex: Penguin Books, Ltd., 1946.

_____. *Odyssey*. Translated by E.V. Rieu. Revised by D.C.H. Rieu and Dr. Peter V. Jones. Middlesex: Penguin Books, Ltd., 1991.

Hymns and Epigrams. Translated by A.W. Mair, Aratus, in Callimachus. London: Loeb Classical Library, 1921.

Kaeppel, C. *Off the Beaten Track in the Classics.* Melbourne: Melbourne University Press, 1936.

Kerenyi, K., *Athene Virgin and Mother in Greek Religion.* Connecticut: Spring Publications, Inc., 1988.

Livy. *Livy Rome and the Mediterranean.* Translated by Henry Bettenson. Middlesex, Penguin Books, Ltd., 1976.

Lyra Graeca II. Translated by J.M. Edmonds. London: Loeb Classical Library, 1924.

Marvell, Andrew. *The Complete English Poems.* Edited by Elizabeth Story Donno. London: Allen Lane, 1974.

Meister Eckhart. *Meister Eckhart,* Vol. I. Edited by Franz Pfeiffer. Translated by C. de B. Evans. London: John M.Watkins, 1956.

Murray, Gilbert. *The Rise of the Greek Epic.* Oxford: Clarendon Press, 1924.

Ovid, Vol. III, *Metamorphoses.* Translated by F. J. Miller. Revised by G.P. Gould. London: Loeb Classical Library, 1977.

Parmenides. Translated by F. M. Cornford. London: Loeb Classical Library, 1961.

Pausanias. Translated by W.H.S. Jones, et al. London: Loeb Classical Library, 1918.

Plato. *Plato: Collected Dialogues.* Edited by Edith Hamilton and Huntington Cairns. Princeton: Princeton University Press, 1963.

————. *Republic.* Translated by Desmond Lee. Middlesex: Penguin Books, Ltd., 1974.

————. *The Dialogues of Plato.* Translated by Benjamin Jowett. London: Encyclopaedia Brittanica, 1952.

————. *Timaeus and Critias.* Translated by H.D.P. Lee. Middlesex: Penguin Books, Ltd., 1965.

Plotinus. *Enneads.* Translated by S. McKenna. London: Faber & Faber Ltd., 1969.

Plutarch. *Plutarch's Moralia.* London: Loeb Classical Library.

————. *Parallel Lives.* London: Loeb Classical Library.

Porphyry. *On Abstinence from Animal Food.* Translated by Thomas Taylor. London: Centaur Press, 1965.

Rumi, Jalaluddin. *The Mathnawi of Jalaluddin Rumi*. Edited by R.A. Nicholson, 1925–1940. Leiden and London.

Seferis, George. *Collected Poems 1924-1955*. Translated and edited by Edmund Keeley and Philip Sherrard. London: Jonathon Cape, 1973.

Severin, Tim. *The Ulysses Voyage*. London: Arrow Books, Ltd., 1987.

Sophocles. *Antigone, Women of Trachis, Philocteles, Oedipus at Colonnus*. Translated by H. Lloyd-Jones. London: Loeb Classical Library, 1994.

———. *Ajax, Electra, Oedipus Tyrannus,Vol. I*. Translated by H. Lloyd-Jones. London: Loeb Classical Library, 1994.

———. *Oedipus the King, Oedipus at Colonus, Antigone, Vol.1*. Translated by F. Storr. London: Loeb Classical Library, 1981.

Strabo. *Geography*. Translated by H.L.Jones. London: Loeb Classical Library, 1969.

Synthese. Edited by Rush Rhees, Vol. 17, 1967, 'Vemerkungen über Frazer's The Golden Bough'.

Taran, Leonardo. *Parmenides*. Princeton: Princeton Univ. Press, 1965.

Taylor, Thomas. *Thomas Taylor the Platonist: Selected Writings*. Edited by Kathleen Raine and George Mills Harper. Princeton: Princeton University Press, 1969.

The Apocryphal New Testament. Translated and edited by J.K. Elliot. Oxford: Clarendon Press, 1993.

The Critic as Anti-Philosopher. Edited by G. Singh. Atlanta, GA: University of Georgia Press, 1982.

The Hymns of the Rg Veda. Edited by R.T.H. Griffith. Benares: 1963.

The Pageant of Greece. Edited by R.W. Livingstone. Oxford: Oxford University Press, 1935.

The Penguin Book of Greek Verse. Edited by Constantine A. Trypanis. Middlesex: Penguin Books, Ltd., 1972.

The Pythagorean Sourcebook and Library. Edited by Kenneth Sylvan Guthrie. Michigan: Phanes Press, 1987.

The Renaissance Philosophy of Man. Edited by E. Casirer et al. Chicago: University of Chicago Press, 1948.

Thucydides. *History of the Peloponnesian War*, Vols. I & II. Translated by Charles Forster Smith. Cambridge, MA: Harvard University Press, 1977.

Virgil. *Aeneid*. Translated by W.F. Jackson-Knight. Middlesex: Penguin

Books, Ltd., 1958.

————. *Aeneid, Books I–VI, Eclogues and Georgics.* Translated by H. Fairclough. Revised by G. P. Gould. Loeb Classical Library, 1916.

Works of Plato, 5 volumes. Translated by Floyd Sydenham and Thomas Taylor. New York: AMS Press, 1979.

Xenophon, IV. *Memorabilia and Oeconomicus, Symposium and Apology.* Translated by E. C. Marchant and O. J. Todd. London: Loeb Classical Library, 1923.

Yeats, William Butler. *A Vision.* London: Macmillan & Company, 1962.

www.ingramcontent.com/pod-product-compliance
Lightning Source LLC
Chambersburg PA
CBHW031559110426

42742CB00036B/249